NEGRO INTELLIGENCE AND SELECTIVE MIGRATION

Prepared under the auspices of the
Columbia University Council for Research
in the Social Sciences

NEGRO INTELLIGENCE

and

SELECTIVE MIGRATION

by

Otto Klineberg

NEW YORK: MORNINGSIDE HEIGHTS

COLUMBIA UNIVERSITY PRESS

M·CM·XXXV

PRINTED IN THE UNITED STATES OF AMERICA

GEORGE BANTA PUBLISHING COMPANY, MENASHA, WISCONSIN

TO

S. R. G.

GRATEFULLY

ACKNOWLEDGMENTS

Acknowledgment is gratefully made to the Columbia University Council for Research in the Social Sciences for the research grant which made this investigation possible; to the departments of Psychology and Anthropology at Columbia University for together sponsoring the project; to the graduate students who engaged in the experimental and statistical studies and whose names are mentioned in the text; to the school authorities in Nashville, New Orleans, Birmingham, Atlanta, Charleston, and New York for permission to make use of the school records and to carry out the testing program in the schools; to the many school principals and teachers who kindly coöperated in this program; to Professor Joseph Peterson of George Peabody College and to Professor Charles S. Johnson of Fisk University for valuable advice in the early stages of the investigation; and to Professor Franz Boas for his constant guidance.

OTTO KLINEBERG

COLUMBIA UNIVERSITY
New York, N.Y.
December 10, 1934

CONTENTS

I. THE PROBLEM 1

II. THE MIGRANTS 6

III. SCHOOL RECORDS 14

IV. NEW YORK CITY—GROUP TESTS 24

V. NEW YORK CITY—STANFORD-BINET TESTS 43

VI. NEW YORK CITY—PERFORMANCE TESTS 47

VII. CITY AND COUNTRY 53

VIII. SCHOOLS NORTH AND SOUTH 56

IX. DISCUSSION 59

BIBLIOGRAPHY 63

INDEX 65

TABLES AND GRAPHS

Tables

1. Northern and Southern Negroes, army results   1
2. Southern Whites and Northern Negroes, by states, army results   2
3. Miscellaneous studies of Negro children, North and South   2
4. School scores of various groups of migrants   21
5. Age scores of various groups of migrants   21
6. Northern migrants classified according to date of migration   22
7. National-Intelligence-Test score and length of residence (Lapidus) . . .   25
8. Combined groups (Lapidus)   26
9. Grade retardation (Lapidus)   27
10. National-Intelligence-Test score and length of residence (Yates)   30
11. Combined groups (Yates)   31
12. Grade retardation (Yates)   32
13. Migrants from city and country (Yates)   32
14. Grade retardation, city-born and country-born groups (Yates)   34
15. National-Intelligence-Test score and length of residence (Marks)   35
16. Results with West Indian and Panama cases omitted (Marks)   36
17. Average grade and grade retardation (Marks)   36
18. Comparison of 1931 and 1932 averages   37
19. Comparison of boys and girls   40
20. The three studies combined (Lapidus, Yates, and Marks)   40
21. Otis scores and length of residence (Traver)   41
22. Stanford-Binet intelligence quotients and length of New York residence, ten-year-old girls (Skladman)   43
23. Stanford-Binet intelligence quotients and length of New York residence, ten-year-old girls (Wallach)   44
24. Stanford-Binet intelligence quotients and length of New York residence, ten-year-old boys (Rogosin)   45
25. Stanford-Binet intelligence quotients and length of New York residence, ten-year-old boys and girls combined   46
26. Pintner-Paterson scores and length of residence (Brown)   47
27. Speed of movement, Healy "A" (Brown)   49
28. Speed of movement, Dotting Test (Brown)   49
29. Paper-form-board scores and length of residence   50

30. Curtis Arithmetic Test scores and length of residence (Horowitz) 52

31. Test score and length of residence in city 54

32. Per capita expenditure on education 57

33. Relation between skin color and length of New York residence, ten-year-old boys (Brown) 60

34. Relation between Negroid characteristics and length of New York residence, twelve-year-old boys (Holmes) 61

Graphs

1. Distribution of school marks obtained by Northern migrants from Birmingham and Nashville 18

2. Distribution of "age scores" obtained by Northern migrants from Birmingham and Nashville 19

3. Relationship between school marks and date of migration, Birmingham migrants 22

4. Relationship between school marks and date of migration, Nashville migrants 23

5. National-Intelligence-Test scores and length of residence in New York, twelve-year-old boys (Lapidus) 26

6. National-Intelligence-Test scores and length of residence, combined groups (Lapidus) 27

7. National-Intelligence-Test scores and length of residence in New York, twelve-year-old girls (Yates) 30

8. National-Intelligence-Test scores and length of residence, combined groups (Yates) 31

9. National-Intelligence-Test scores and length of residence, city and country-born (Yates) 33

10. Comparison of results obtained in 1931 and 1932 (Lapidus and Marks) . . 38

11. Comparison of boys and girls (Marks and Yates) 39

12. Otis scores and length of residence, twelve-year-old girls (Traver) 41

13. Pintner-Paterson mental age and length of residence in New York (Brown) 48

14. Minnesota-Form-Board scores and length of residence (Horowitz) 50

15. Curtis Arithmetic scores and length of residence (Horowitz) 51

NEGRO INTELLIGENCE
AND SELECTIVE MIGRATION

CHAPTER I

THE PROBLEM

Since the days of the Army intelligence-testing program a very large amount of material dealing with the question of Negro intelligence has been collected. The summaries of the results by Garth (9*), Pintner (21), Witty and Lehman (26) and others make it quite clear that Negroes rank below Whites in almost all studies made with intelligence tests. An analysis of these results soon shows, however, that the amount of difference between the two groups varies very considerably from one part of the country to another. More specifically, northern Negroes do very much better in the tests than Negroes in the South, and approximate much more closely the records made by the Whites with whom they are compared.

This difference between northern and southern Negroes was first clearly demonstrated in the results obtained during the War (17). One comparison between 14,994 southern, and 8,165 northern Negroes gave the following percentage distribution of letter grades:

TABLE I

NORTHERN AND SOUTHERN NEGROES, ARMY RESULTS

	No.	D−	D	C−	C	C+	B	A
Northern Negroes	8,165	19.6	27.6	22.1	21.4	6.7	2.3	0.6
Southern Negroes	14,994	55.7	26.4	9.8	6.2	1.4	0.4	0.1

Although the northern Negroes still rank below the northern Whites, they are clearly superior to the larger group of Negroes from the South. It is well known that the Negroes in certain of the northern states actually exceeded the median scores obtained by White recruits from a number of the southern states, as shown in Table 2.

Very few of the studies made since the War have been directly concerned with the comparison of scores made by Negroes in different parts of the country. It is interesting, however, to list separately the studies made in the North and in the South to see whether the differ-

* The numbers in parentheses refer to the Bibliography on p. 63.

TABLE 2

SOUTHERN WHITES AND NORTHERN NEGROES, BY STATES, ARMY RESULTS

Whites		*Negroes*	
State	*Median Score*	*State*	*Median Score*
Mississippi	41.25	Pennsylvania	42.00
Kentucky	41.50	New York	45.02
Arkansas	41.55	Illinois	47.35
Georgia	42.12	Ohio	49.50

ence found by the Army testers in the case of recruits also holds for the more recent studies of Negro children. In the following table those studies are listed in which an I.Q. (intelligence quotient) for Negro children is given or can be calculated from the results, in order that the findings may be combined in such a way as to make a comparison between North and South possible. This procedure suffers from the difficulty that intelligence quotients obtained by different investigators and

TABLE 3

MISCELLANEOUS STUDIES OF NEGRO CHILDREN, NORTH AND SOUTH

Investigator	*State*	*Test*	*I.Q.*
North			
Arlitt	Pennsylvania	Binet	83.4
Barnes	Kansas	Binet	84.6
Goodenough	California	Goodenough	85.8
Kempf, Collins	Illinois	Various tests	71.0
Lacy	Illinois	Binet	85.6
Phillips	Pennsylvania	Binet	93.0
Pintner, Keller	Ohio	Binet	88.0
Pressey, Teter	Indiana	Pressey	83.0
Schwegler, Winn	Kansas	Binet	89.2
Strachan	Missouri	Binet	93.4
Strachan	Missouri	Pintner-Cunningham	92.0
South			
Davis	Texas	Terman	78.0
Garth	Southern States	Otis	77.0
Garth, Whatley	Texas	National Intelligence Test	75.0
Goodenough	Southern States	Goodenough	78.7
Graham	Georgia	Binet	99.0
Hirsch	Tennessee	Pintner-Cunningham	84.6
Jordan	Arkansas	National Intelligence Test	75.0
Lacy	Oklahoma	Binet and Otis	91.0
O'Shea	Mississippi	Various tests	75.0
Peterson	Tennessee	Pressey	75.0
Peterson	Tennessee	Otis	58.0
Peterson	Tennessee	Haggerty	92.0
Sunne	Louisiana	Binet	78.0

with different tests are not directly comparable; the averages obtained for North and South should therefore not be taken too strictly, but rather as a general indication of the trend of the studies (21, 9, 26).

The average I.Q. for all the northern groups is 86.3; for the southern groups, 79.6. The Negro children, however, both northern and southern, do somewhat better in the Binet tests than in the group tests, and, since a large number of the northern studies were conducted by means of the Binet, that fact might account for at least part of the difference. When averages are separately calculated for the Binet results and the results obtained by group tests, the difference is still definitely in favor of the northern children for the latter, the average I.Q. being 82.9 in the North, and 76.8 in the South. With the Binet there is practically no difference, the figures being 88.2 and 88.5, respectively.

In the interesting study by Peterson and Lanier (19) it is suggested that "a useful check upon the reliability of a given race difference obtained in any locality and under any specific set of circumstances is to take what seems to be fairly representative samplings from widely different environments and to compare the various results as checks upon one another with a view to determining just which factors persistently yield differences in favor of one or the other race." With this in mind these investigators gave a number of tests including a Binet group test, the Myers Mental Measure, five of the International group rotator tests, and three Ingenuity tests devised by Peterson, to twelve-year-old White and Negro boys in Nashville, Chicago and New York. Not all the tests were given to all the groups, but it was still possible to make a large number of intra as well as interracial comparisons. In general it may be said that while the Whites in Nashville definitely and reliably surpassed the scores made by the Negroes, the difference in favor of the Whites in Chicago was not nearly so marked, and in New York none of the differences between Whites and Negroes was reliable. The results give definite indication, therefore, of a marked difference between northern and southern Negroes, as well as of a clear tendency for northern Negroes—at least in New York—to approach very closely the results obtained by the Whites.

These results are exceedingly important, for on their interpretation probably depends the final decision as to whether there are fundamental differences between Whites and Negroes in the ability to solve the problems presented by tests of intelligence. Two explanations have been suggested. It has been pointed out that there are such marked differences between the environmental opportunities of northern and southern Negroes—differences in expenditures for schooling, in extracurricular activities, in the chance to acquire an education in the wider sense—that these might easily account for the superiority of northern Negroes. It is quite generally admitted at the present time that most if not all intelligence tests are at least in part dependent upon educational and cultural background; and it is certainly conceivable that the superior showing of the northern Negroes may be determined by their more stimulating environment. On the other hand, it has been urged that in migrations of Negroes from South to North definite selective factors have been at work, causing the more intelligent stocks to leave and the less intelligent to remain behind. In that case the Negroes now in the North would not represent an average group obtaining high scores because of the better environment, but a group that was superior to start with. If there has been a selective migration of the more intelligent Negroes, then the differences between North and South do not necessarily point to a definite environmental effect upon Negro intelligence-test scores. If there has been no selective migration, then the differences between North and South can be explained only in terms of such an environmental effect, and they indicate that Negro "intelligence" can markedly improve when there has been a corresponding improvement in the environment. Further, since in the North the Negro scores (as in Peterson and Lanier's study) approach very closely those obtained by the Whites, there will be no basis for the assumption of any thoroughgoing or fundamental White superiority in "intelligence." It seems to the writer, therefore, that the important problem in this field is to determine whether or not there has been a selective migration of Negroes from South to North, and whether such a selection can account for the observed differences.

It is then the problem of selective migration with which the present monograph is concerned. The attempt has been made to develop certain objective methods for the study of the problem. Most writers on this topic have been inclined to a somewhat deductive and a-priori approach. Those who believe in selective factors in migration point out that it requires energy and initiative to start over again in a new setting, as well as intelligence enough to see the advantages of the new environment over the old. It has also been argued, however, that those who are more successful in the old environment, who have achieved a certain social and economic position, who possess property and friends, are less likely to wander off in search of fresh opportunities than those who are shiftless and unsuccessful and have nothing to lose by leaving. Interestingly enough, people in the South with whom the writer discussed this question were inclined to give either the one or the other of these extreme views; it was rare to hear the opinion that those who left were neither better nor worse than those who stayed behind. In any case, nothing of any scientific value can be obtained from this type of subjective generalization.

As an introduction to the more quantitative aspects of the present study, the next chapter reviews briefly the more important social and economic factors which have probably entered into Negro migrations in this country. An examination of old newspaper files very kindly placed at the disposal of the writer by Dr. Monroe C. Work of Tuskegee Institute yielded some interesting contemporary side lights on the character of the migrants, which it has seemed worth while to reproduce in some detail. There is also some direct information, obtained from the migrants themselves or from their families, which is pertinent to the problem of why certain people decide to leave and others decide to remain at home.

CHAPTER II

THE MIGRANTS

There has always been some movement of Negroes from South to North. Before the Civil War a great many slaves made their escape, for example, through the famous underground railroad, and settled permanently north of the Mason and Dixon line. In 1879 there was a large emigration from the southern plantations toward the Middle West, and more particularly to Kansas. In 1888 and 1889 there was a similar migration to Texas and the neighboring states. Migration continued to some extent during the succeeding years, and by 1910 there were over a million Negroes in the northern states.

The great migrations to the North took place, however, immediately before and during the World War. Between 1915 and 1918 the movement of Negroes to the North became so marked and so noticeable as to arouse discussion in the press, both Negro and White, in almost all sections of the country. There were a great many social and economic causes which contributed to this movement, probably the most important being the need for unskilled labor in the North owing to decreased European immigration and to recruiting for the War; the boll-weevil plague in 1915-16, which added to an industrial depression already very marked in the South; the promise, held out by such northern Negro papers as the *Chicago Defender*, of high wages and a better standard of living in the North; and a general dissatisfaction on the part of the southern Negro. The increased demand for labor in the coal and iron mines, steel foundries, factories and stockyards led to the sending of labor organizers to the South, who recruited Negro laborers for their respective industries and supplied them with free transportation to the North. Migration was in this way stimulated by forces operating in northern as well as in southern states.

At the close of every census period statistics of the birthplace and residence of Negroes indicated a steady movement from one section to another within the South, and from rural to urban areas, as well as from South to North (Kennedy, 14). According to the 1920 census the proportion of Negroes who had moved from one state to another was larger than in any previous decade. In 1920, 19.9 percent of Negroes born in the United States were living in a state other than that of their birth (*ibid.*). Most of this migration was into a relatively small number of large northern cities. Between 1910 and 1920 the Negro population of Detroit, for example, increased 611 percent; in Cleveland the increase was 307 percent, in Chicago, 148 percent, and there was more than a 50 percent increase in New York, Philadelphia and St. Louis (1, 27).

The migration continued after 1920, reaching another high point in 1923 and 1924. In 1923 it was estimated that the number of colored children in northern schools had increased over 50 percent since the beginning of the migration (Kennedy, *ibid.*). Figures obtained by the Detroit school census in 1925 show an increase of over 100 percent in the number of Negroes since 1920. It was estimated by the National Urban League (Wesley, 25) that the volume of increase in the Negro population of New Jersey and New York between 1921 and 1925 was two-thirds as great as during the whole decade from 1910 to 1920. Since 1924, migration has continued, but at a considerably lower rate, and during the recent depression years (1931-34) there has even been something of a return movement to the South.

The description of the migrants in the newspaper reports of 1916-17 and 1923-24, two periods in which migration was proceeding most vigorously, differs so much from one writer to another that it is quite impossible to learn from them how the migrants really compared with the general population. Unfortunately, most of the newspapers had a definite ax to grind. The Negro papers, both North and South, were usually anxious to prove to the southern Whites that they were losing their best Negroes as a result of the oppressive measures which they had adopted; their descriptions of the migrants usually refer to them in rather glowing terms. The southern White papers, on the other hand, sought in a great many instances to allay any fears as to the possible loss of a valuable working element by describing the migrants as shiftless and incapable, and as constituting the most undesirable part of the Negro population. The northern White papers could usually be relied upon for a more objective comparison between the newly-arrived southern Negroes and those who had been living in the North for some time. In their case, too, however, the element of bias was by no means absent; as the Whites in the northern cities viewed with considerable alarm the ever-increasing proportions of the migration, they tended to be somewhat hostile to the newcomers and to see in them a definitely inferior group. There were of course many exceptions in all these groups.

A few quotations from these newspaper accounts will illustrate the wide diversity in the opinions expressed.

Philadelphia Christian Recorder, Negro, August 3, 1916:
Of course they are not the best Negroes that leave the South. For the best Negroes do not need jobs as railroad or section hands.

Philadelphia Times, White, July 25, 1917, article by Edgar Mels:
Philadelphia is face to face with a serious problem . . . the problem of the thousands of uneducated Negroes who are pouring from the South into the Northland. . . . The ignorance of the emigrant is amazing.

Nashville Banner, White, March 2, 1917:
The Negroes who are being carried North are wholly of the labor class,—unskilled labor. They are the untutored "field-hands" or river "rousters" and are wanted only for their brawn.

Boston Transcript, White, March 9, 1917:
It takes some enterprise and resolution for a Negro to emigrate from the South to the North; those who come have probably for the most part made up their minds to struggle and adapt themselves.

Chicago Tribune, White, July 10, 1917:
The present invasion of colored people from the South is made up chiefly of discontented tenant farmers and unsettled young men from the farms and small towns. They do not fairly represent the better class of Southern Negroes, hundreds of thousands of whom own their little farms. . . . Those people are not likely to come North. They like the South and its climate and they are intelligent enough to know that they are more independent and better-off in their old homes than they can hope to be in any Northern city. . . . It is very largely the illiterate and penniless Negro who has come North in the last two years.

Birmingham Age-Herald, White, March 21, 1917:
Too many people have the idea that it is the alley Negro, the shiftless class, that are discontented and leaving. This is a great error. Many of the very best among us are leaving, and others are in a state of unrest which we hope our White friends will help to compose instead of fomenting.

Louisville Courier Journal, White, April 1, 1917:
Assuming that the more energetic and ambitious among Southern Negroes are going to the North, the South will suffer the loss of useful labor.

Dallas Express, Negro, March 21, 1917:

This exodus is by no means confined to the worthless or ignorant Negro. A large percent of the young Negroes in this exodus are rather intelligent.

Portland Express (Me.), White, March 23, 1917:

The Negro as the North knows him is educated, polished, able and a keen competitor of the White in every profession and occupation. The Negro who is coming now is the still uneducated laborer, for the most part unskilled, and in the bulk the chief component of the Negro problem as we hear of it from the South.

Some of the comments on the situation arising in the northern schools make it quite clear that as far as previous education is concerned, the southern Negro child was greatly retarded, and presented a very real problem in connection with his assimilation into the school system.

Hartford Post, White, September 15, 1917:

One difficulty in the public schools is that the Negro children from the South, having lacked scholastic training, are far behind White children, hence are much older than the Whites in whose classes they are placed, which creates an awkward situation.

Philadelphia Christian Recorder, Negro, October 6, 1917:

The hundreds of thousands of our people who come here bring with them a great number of children, and these create an educational problem. . . . We find a large number of children who never went to school and yet are of school age. We find also a large number of large-sized children, ten and twelve years of age, who must enter first and second grade with children of from five to seven years of age. We find a large number of children mentally dull because they have not been to school before or have had no competent teaching.

Cincinnati Inquirer, White, September 12, 1917:

Mr. Roberts (Assistant Superintendent of Schools) said that . . . there was the trying problem of dealing with and providing educational facilities for children, many of whom are greatly retarded in their studies. . . . Some of the children entering the city schools ranging in age from nine to fourteen years barely have learned to read and write. Others cannot do that, it is said.

The newspaper accounts of 1923 and 1924 usually describe the migrant in rather more favorable terms. There seems to be a general tendency to regard the exodus as having first affected mainly discontented and unadjusted persons, but as having spread gradually to other classes until it involved a fair cross section of the southern Negro population.

Philadelphia North American, Negro, March 6, 1924:

According to the annual report of the Armstrong Association (Forrester B. Washington) . . . there have been two main types in this migration. In 1917-18 the War demand brought the floater, the unattached irresponsible hunter of temporary jobs. . . . Since 1918 a more substantial type has come, including some mechanics and professional men, also farmers and business men, but consisting in the main of unskilled laborers who differ from the earlier comers chiefly in that they are of the steadier type.

Birmingham Reporter, Negro, January 13, 1923:

Observation and testimony bear witness that those going and those planning to go are not the 'riff-raff,' not the thriftless, not the most ignorant, but a large percentage of prosperous, industrious and intelligent Negro citizens.

Pittsburgh Courier, Negro, September 29, 1923:

The migration of Negro laborers to Northern industries has been accompanied by an exodus of colored business and professional men. Preachers have had to move to Chicago, Detroit or Cleveland to keep up with their congregations. One church in a nearby town had three hundred members a year ago. Today it has only one, an ancient unable to stand the railroad trip. Last week the preacher packed his belongings and joined the procession. It is cited by leaders as a typical case. . . . The present wave of migration, unlike that of the War period, is mainly composed of the superior type of Negro laborer.

Opportunity, Negro, October, 1924, article by George E. Haynes on "Negro Migration":

Frequently whole families or neighborhoods, sometimes with previous arrangement for employment in some of the industrial communities, migrated in a group. A few cases have been recorded of whole church congregations bringing their pastors with them.

In spite of the diversity of opinions expressed, a comparison between earlier and later accounts makes it probable that the first migrants left the South largely as a result of the local activities of northern labor organizers. These organizers were not particularly careful in their selection of men, and the first movement (1917-18) undoubtedly included many of the less desirable members of the southern Negro population. Later the movement spread to others. Woofter (27) writes:

Labor agents canvassed the South for men, and coming on the heels of the disorganization of cotton farming caused by the boll weevil they secured large numbers. Others followed without solicitation upon hearing of the success of the first migrants. Also the boll weevil ravaged hundreds of thousands of acres of fertile cotton fields, and Southern farmers passed through a period of depression. Many farmers told their laborers to leave. Negroes began to discern the poverty of educational facilities and the insecurity of life in certain rural communities.

A detailed study by Louis E. King (unpublished dissertation for the degree of Doctor of Philosophy, Department of Anthropology, Columbia University) of 110 migrants from rural communities in West Virginia to northern cities gives interesting information as to some of the factors which may be operative. By questioning either the migrants themselves or their families, King was able to discover exactly why many of them had left. Of the 56 male migrants, 16 went North in order to make better wages—intelligence may have played a part here—but 12 left because they could find no employment at home, 9 went as a result of direct inducement on the part of relatives and friends in the North, 2 because of misdemeanors which they had committed, and 2 were taken North by a labor agent. Four were sent to the city by their parents to go to school, but out of all the 56 male migrants only 2 had any schooling after leaving their homes. The women migrants left for very similar reasons. Fifteen went North because they could find no work at home, 14 in order to make better wages, 6 were invited by friends and relatives and stayed on, 5 got married and accompanied their husbands, 4 were taken to the cities by their employers, and 2 went in order to continue their schooling. Of the 54 female migrants 4 had some subsequent schooling.

Perhaps the most interesting result of King's inquiry is the discovery that most of those who left had definite connections with someone who had previously migrated to the city or to the North. Fifty-nine migrants had relatives and 29 others had friends in the cities to which they moved, and in almost every case it was a suggestion or an invitation from one of them that prompted the migration. This being the case, the "energy and initiative" which has so often been regarded as a prerequisite to migration need hardly play an important rôle. King is of the opinion that those who migrate do not stand out in any special way from those who remain at home, and that their departure is to be understood in the light of very specific and frequently accidental factors.

Peterson and Lanier (19) in the monograph quoted in the preceding section base their argument for selective migration upon the proportion of Negroes to the total population in Tennessee, Illinois and New York. They believe that the decreasing difference between Negroes and Whites in Nashville, Chicago and New York City is probably a function of the increasing severity of selection of Negroes, as indicated by the decreasing percentage of their population. This argument does not seem very convincing to the writer in view of the very large Negro populations in Chicago and New York. The southern Negro migrant to these cities finds there a large thriving Negro community, almost self-sufficient in its economic life, and with social, economic and educational agencies far superior to those in most southern cities. His struggle for existence is certainly not any more severe than, for example, in Nashville or New Orleans. It is hard to see, therefore, why the decision to go to New York or to Chicago, rather than elsewhere, should argue for more rigorous selection in the case of these particular Negroes. In any event, it is difficult to see why the proportion of Negroes in the three states, rather than the total number in each of the three cities should be taken as a criterion of the rigor of the selection. New York is the largest, and Nashville the smallest, of the three Negro communities. As on the whole the Negroes do constitute more or less self-contained groups, it seems even more reasonable to use the absolute rather than the proportionate figures, in which case the argument for the selective nature of migration to New York City would hardly hold.

In general it is fair to say that the problem of selective migration cannot be settled by arguments of a purely logical nature; such arguments seem to lend themselves equally to the support of both sides of the issue. In the present study an attempt has been made to approach the problem as objectively as possible. Two methods were used. In the first place it was felt that, if at all possible, some direct comparison should be made between the intellectual level of those who left the South and those who stayed behind. For this purpose the school records in three southern cities—Nashville, Tennessee; Birmingham, Alabama, and Charleston, South Carolina—were examined for information as to whether those Negro children who had migrated to the North represented in their school work a superior or an inferior group when compared with the total Negro school population in those cities. School records are admittedly defective as measures of intelligence, but they were the only measures available in this study. A second and more indirect method was to give intelligence tests to southern born Negro children now living in New York City, but differing in the number of years of residence there, in order to determine whether there is any noticeable change in their scores proportionate to the length of time they have been in New York. If the difference between northern and southern Negroes is entirely due to selective factors, length of residence should make no appreciable difference; if the environmental factors are important, there should be an improvement in direct relation to the length of time such factors have been operative.

The following chapters present the results of these studies in detail.

CHAPTER III

SCHOOL RECORDS

In this portion of the study the problem is to see how the school records of the migrants before they migrated compared with the records of those who remained in the South. In spite of the admitted defects of school marks as an indication of intelligence level, it was felt that for lack of something better they might be used to advantage. It would, of course, have been preferable to use intelligence-test records, but there were very few cases of southern Negro schools in which intelligence tests had been in use for more than the past year or two, so that the number of migrants for whom an intelligence quotient was available was too small to be of significance. In any case, since the intelligence tests in general use have almost always been standardized against school records, it was felt that the latter, in spite of their rather arbitrary character and lack of objectivity, might serve as a useful criterion.

Through the kindness of the school authorities in Nashville and Birmingham, the school registers over a period of years were examined for cases of children who had left for one or another of the northern cities. In Nashville the records studied covered the years from 1921 to 1930 inclusive, and in Birmingham, from 1914 to 1924 inclusive. After 1924 the Birmingham school authorities instituted a new system of grading the children, substituting letter grades for actual percentages. As a statistical evaluation of the results in terms of letter grades would not have been very satisfactory, it was decided to restrict this investigation to the school years for which a more quantitative estimate was available. A supplementary group of records from Charleston, South Carolina, was later collected.

Certain precautions were observed in order to make the records as objective as possible. If the register showed that in 1919 John Jones in the fifth grade of a Nashville public school had left for Detroit or Cincinnati, the mark with which he was credited was that obtained in the last school grade completed, in this case the fourth, so that the mark could not be affected by departure in the middle of the school term. In addition, a mark of 60 percent obtained in one school from one teacher in, let us say, 1919, was not assumed to be identical with a similar mark obtained from another teacher in another school in 1923. As it was necessary to have some common basis of marking, it was decided to throw the records for every class into a rank-order distribution, and the mark obtained by the migrant was determined by his position in this rank order. In other words, the migrant's record was a function of his position in the class and of the number of children in that class. The formula used was percent position $= \dfrac{100(R-.5)}{N}$, in which R equals rank and N equals number of cases. Clark Hull's tables were used for transmuting percent position into rank score. By this method a mark of 50 would signify that the migrating child was exactly at the average for his class (8).

The school register did not always contain information as to the destination of the migrant. In many cases the statement was simply "left school," "moved," or something equally indefinite. In such cases the teachers or the principal of the school could usually supply the missing information and, in still other cases, it was possible to discover from other Negro families in the neighborhood just where the migrant had gone. In this part of the study two Negro assistants, Miss Annabelle Bloodgood of Fisk University in Nashville, and Mrs. F. Carter, substitute school teacher in Birmingham, were able to obtain useful information which otherwise might not have been easily accessible. This procedure still left a number of migrants whose destination could not be discovered; their records were of course not included. In those cases in which teachers, principals or neighbors had simply a general idea of the present location of the migrant, for example, that he went "North" or "out to the country," it was also decided not to use the records. The material as here presented, therefore, includes only those migrants for whom a *specific* destination could be ascertained, and while certain errors may have crept in as the result of later migration, it is felt that on the whole the information received may be regarded as accurate.

The question will naturally arise as to the extent to which records of school children may be considered representative of the migrating population. These children were not in any sense the originators of the migration; they went passively with their parents. May it not be that even though the adult migrants constituted a superior, selected group, their children would not reflect any such superiority? This is of course a possibility, and it would have been preferable to have collected material relative to the intelligence of the migrants themselves rather than that of their children. This was unfortunately not possible. For the implications of this particular study, however, the records of the children may be regarded as equally satisfactory. Contained within the hypothesis of selective migration as explaining the differences between northern and southern Negroes, there are implicit two distinct hypotheses—first, that more intelligent people migrate; and second, that these "superior" migrants have "superior" children. When northern and southern Negro children are compared, and the superiority of the former is explained as a result of selective migration, it is clearly a selection of the parents rather than of the children which is assumed. The children now in New York City, for example, obviously did not migrate of their own free will; they were brought North by their parents; if there were any selective factors they operated not upon the children but upon their parents. If, therefore, selective migration of families or stocks is a fact, the school records collected in this study should reveal a higher level of intelligence in the case of the migrating children who went passively, as clearly as in the case of adult migrants who went on their own initiative.

The fact that the hypothesis of selective migration involves the further hypothesis that superiority is inherited, does not imply that the writer is prepared to make this latter assumption. There is probably a certain amount of mental similarity between various members of the same family, but the extent of this similarity is still an open question. It is not likely that the resemblance between parents and children is so close as the hypothesis of selective migration would demand. For the purposes of argument, however, the writer is prepared to grant that mental inheritance may take place in this manner, and to let the present issue rest upon an answer to the more direct question of whether or not it is the more intelligent individuals who migrate.

The statistical computations for this part of the study were made by Miss Frances L. Hand, graduate student in the Department of Psychology at Columbia University.

There were 303 cases of northern migrants from Birmingham. The average score obtained was 44.8 (with a standard deviation of 19.5), that is to say, definitely below the average for the whole population, which is 50. The reliability of the difference between the migrating and non-migrating groups could not be ascertained, as the average score of the latter by this statistical method depends upon the assumption of an infinitely large distribution. In any case there is no evidence in the Birmingham results of a selective migration of the more intelligent children.

In the case of school children the question of age is obviously important. A mark of 75 obtained by a ten-year old boy in the fifth grade clearly means more than the same mark in the case of a fifteen-year-old boy. In the present study this factor was checked by taking note of the age of each migrating child relative to the ages of all other children in the same class. The children in each class were ranked in order of merit for age, that is, the youngest child at the top, and the oldest at the bottom of the list. The migrating child was given a score, based upon percent position as above; if he received a mark of fifty he was exactly at the average age for the class; if above fifty, younger than the average, if below fifty, older. Age may by this method be regarded as a measure of educational achievement, and consequently as a very rough index of intelligence; it is therefore a check upon the results obtained by the use of school marks. The results indicate that the migrating group as a whole was rather older *for its grade* than the rest of the school population. The 308 cases had an average score of 44.9 with a standard deviation of 17.6.

The results in the case of the Nashville migrants to the North show these children to be slightly superior to the average, both in school records and in age rankings. There were 184 cases; the average score for school marks was 54.0 with a sigma of 20.2; the age score was 53.0 with a sigma of 15.4.

When the Birmingham and Nashville results were combined in one larger distribution of 487 cases, the average score of the migrants was very slightly below that of the general population. The accompanying histograms present in detail the results for the combined group.

It was thought that the distribution of scores for the total group might yield some qualitative information as to the nature of the migrants. The opinions expressed in the South as to what sort of people migrated tended definitely toward one extreme or the other; the migrant was judged either as very superior or very inferior, rather than as average. The question arose as to the possibility that the migrants represented the two extremes of the population, so that the distribution curve might present some evidence of bimodality. Beyond the suggestion of some flattening of the top, more marked in the case of the age scores, the curves show nothing unusual. The conclusion seems justified that as far as these results go, they indicate that the migrants constitute an average group containing good, bad and indifferent members of the

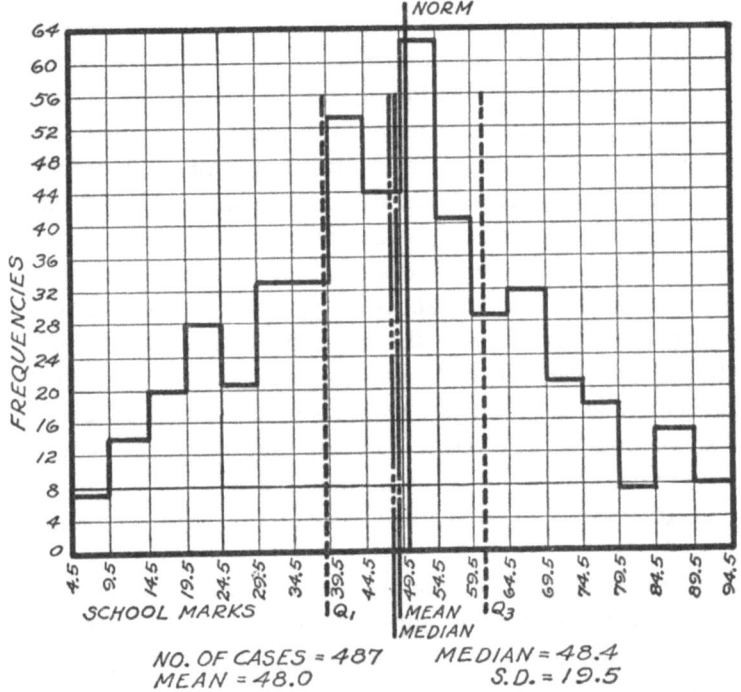

NO. OF CASES = 487   MEDIAN = 48.4
MEAN = 48.0   S.D. = 19.5

GRAPH I. DISTRIBUTION OF SCHOOL MARKS OBTAINED BY NORTHERN MIGRANTS FROM BIRMINGHAM AND NASHVILLE

community. There is no evidence in the group as a whole for the operation of any factors selecting either the more or the less intelligent members of the community.

When the Charleston migrants were included, the total group of 562 cases had an average score of 49.3, almost exactly at the average for the whole population.

The supplementary group of 75 migrants from Charleston, S.C., for the years 1924-30 showed by the same method a median score of 55.6 (S.D. 20.2), that is, somewhat above the average for the non-migrating children. The higher score of the Charleston migrants, and the relatively low score of the Birmingham group, suggest that there may be different factors operative in different communities, so that from one the superior individuals, from another the inferior, will tend to migrate. The reason for this is probably to be found in economic factors; it may be, for example, that where economic conditions are very bad, as they have

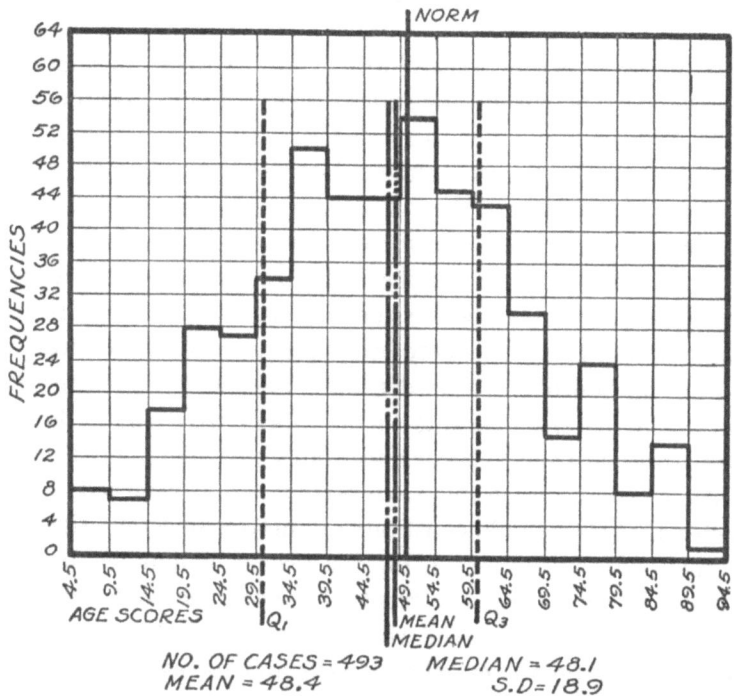

NO. OF CASES = 493   MEDIAN = 48.1
MEAN = 48.4   S.D = 18.9

GRAPH 2. DISTRIBUTION OF "AGE SCORES" OBTAINED BY NORTH-
ERN MIGRANTS FROM BIRMINGHAM AND NASHVILLE

been for some time in South Carolina, those with more enterprise and intelligence try their luck elsewhere; where conditions are better, as they undoubtedly have been in Birmingham, it may be that there is enough work for the more capable ones, and that it is those with less ability who find it to their advantage to leave. This is of course merely a hypothesis, which should be tested by careful investigation of the economic status of the migrants and their relation to the general population. The problem is complicated in the case of our results by the fact that the Birmingham migrants on the whole came earlier than those from Nashville, and that these in turn came earlier than the ones from Charleston. As there is some evidence (see below) that the quality of the migrants has somewhat improved in recent years, it may be that part of the difference in the records for the three cities is due to this improvement, rather than to economic causes. In any case a combined study of the same migrants from both a psychological and an economic viewpoint would be desirable; it was unfortunately not possible in this case.

Not all the migrants went North. Some went to other cities in the South, some to the border-line states (West Virginia, Kentucky), and a not inconsiderable number to rural districts in the vicinity. If the superiority of northern over southern Negroes, which we are examining, is due to selective factors, the migrants to the North ought to be superior to those who have gone elsewhere. The results are conflicting in the case of the two cities (Nashville and Birmingham) from which there were enough migrants to be classified in this manner. From Birmingham the migrants to the North were below the average, while those to other parts of the South or to border-line states were above the average. From Nashville the migrants to the North were above the average, those to the South and to border-line states being definitely below. There was also a group of migrants from Nashville who returned; it might be expected that they would constitute an inferior selection, since presumably they were unsuccessful, for one reason or another, in their new locations. These returned migrants, however, received a mark almost exactly at the average for the whole population. Tables 4 and 5 present these results in detail for school marks and for age scores; it is clear that there is not enough consistency to warrant any conclusion as to an intellectual difference between groups migrating to various parts of the country.

An attempt was made to answer still one more question on the basis of this material. Has selection been increasingly or decreasingly severe during recent years? It is sometimes urged that it is only the first migrants, the "pioneers," who really show any unusual ability; they are the ones who have to create for themselves new homes and new opportunities for earning their livelihood; those who come later presumably benefit by the experience of the former, and are helped and encouraged by them when they arrive. On this basis it might be expected that the quality of the migrants should become poorer as time goes on.

On the other hand, the newspaper descriptions cited in Chapter II suggest that the first migrants were the misfits and the undesirables, and that the more respectable and successful members of the group came later. The school records in this study were examined for evidence on

TABLE 4

SCHOOL SCORES OF VARIOUS GROUPS OF MIGRANTS

Destination of Migrants		No. of Cases	Median
From Nashville	North	184	52.9
	South	110	44.5
	Border-Line States	23	42.0
	Returned	89	50.7
From Birmingham	North	303	44.8
	South	57	52.8
	Border-Line States	20	54.5

this point, and the averages for each successive two-year period from 1914-24 in the case of Birmingham, and from 1921-30 in the case of Nashville, were compared. The results are shown in the accompanying tables and graphs. They indicate that the quality of the migrants has been steadily improving; the number of cases at each interval is not very large, but the results are consistent enough to be suggestive. It will be seen that the only exception is in the case of the very recent migrants from Nashville, who fall slightly below the preceding group, but still remain above the general average. It is interesting to note that the

TABLE 5

AGE SCORES OF VARIOUS GROUPS OF MIGRANTS

Destination of Migrants		No. of Cases	Median
From Nashville	North	183	51.3
	South	107	48.0
	Border-Line States	22	37.0
	Returned	85	52.1
From Birmingham	North	308	48.0
	South	57	42.0
	Border-Line States	22	46.5

Nashville migrants, who on the whole represent more recent arrivals, appear to carry on the upward trend which is already quite noticeable in the successive groups of northern migrants from Birmingham. It is perhaps hardly necessary to point out that local circumstances such as

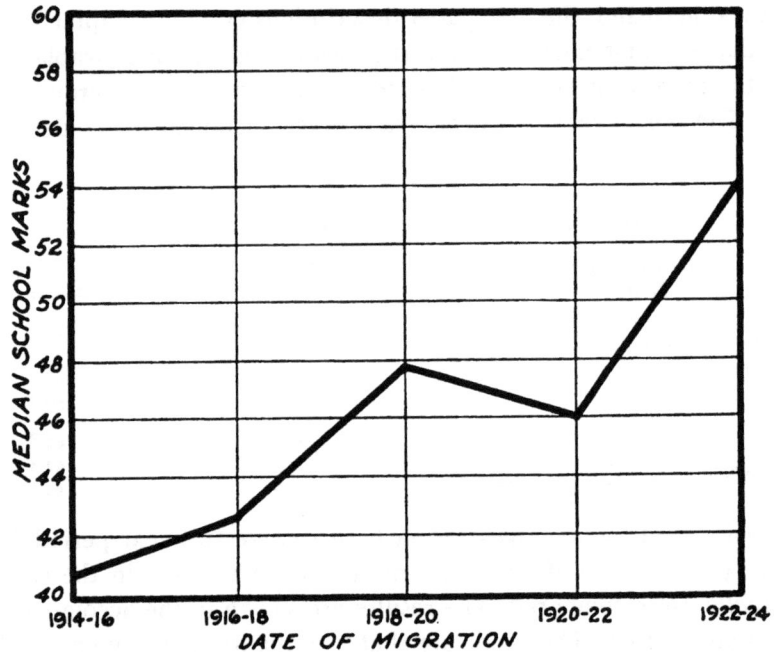

GRAPH 3. RELATIONSHIP BETWEEN SCHOOL MARKS AND DATE OF
MIGRATION, BIRMINGHAM MIGRANTS

the improvement of schooling in the South cannot account for these
results, since in each case the school marks of the migrants were compared with the marks received by non-migrants *in the same year;* it is
the *relative,* not the absolute, school mark which is taken as the measure
of scholastic ability.

TABLE 6

NORTHERN MIGRANTS CLASSIFIED ACCORDING TO DATE OF MIGRATION

	Years	*No. of Cases*	*Median*
	1914-16	61	40.7
	1916-18	126	42.5
From	1918-20	45	47.8
Birmingham	1920-22	33	45.7
	1922-24	40	53.8
	1921-23	33	45.8
From	1923-25	43	52.5
Nashville	1925-27	50	59.1
	1927-30	61	53.6

In summary it may be said that the school records examined give no
evidence that Negro migrants to the North are superior to the non-

migrants, or to those who migrate to other parts of the country. They seem to constitute about an average group. There are differences between the migrants from the three southern cities studied; these may be due to differing economic factors, or to a change in the quality of the migrants in recent years. In general there seems to have been a gradual improvement in the intellectual level of the migrants from 1915 to 1930.

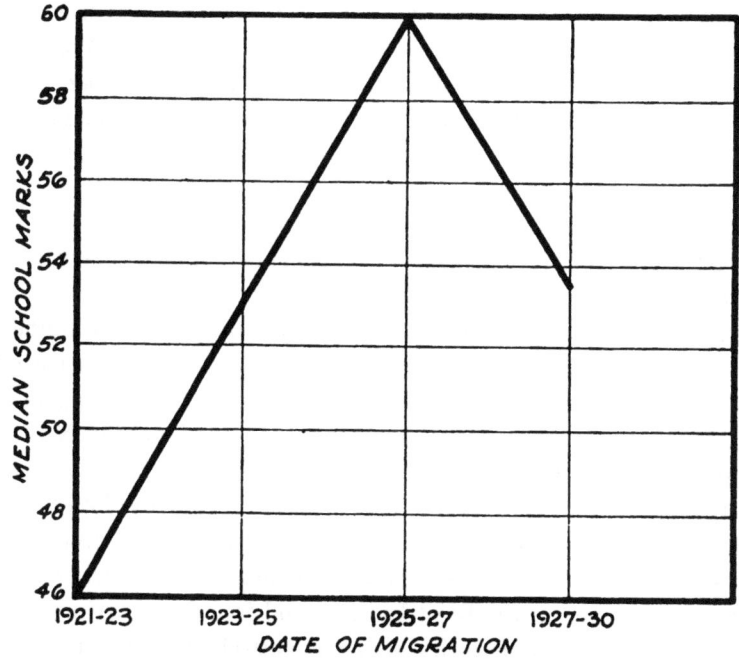

GRAPH 4. RELATIONSHIP BETWEEN SCHOOL MARKS AND DATE OF MIGRATION, NASHVILLE MIGRANTS

This part of the study, therefore, gives no indication of a selective migration which might explain the superiority of northern over southern Negroes. There remains the problem of deciding whether a change in the environment of the Negro children could raise their intelligence-test level sufficiently to account for the observed difference. This problem was attacked directly by a study of southern born Negro children now living in New York City; the results are presented in the following chapters.

CHAPTER IV

NEW YORK CITY—GROUP TESTS

This part of the study attempts to discover whether the admittedly superior northern environment has any effect in raising the intelligence-test scores of southern born Negro children. The method used was to compare the scores obtained by different groups of New York Negro children, all born in the South, but differing in the number of years which they had lived in New York City. If the environment has an effect, there should be a rise in intelligence at least roughly proportionate to length of residence in New York. If there is no environmental effect, and if the superiority of the New York City Negroes is entirely due to selective migration, length of residence ought to make little or no difference.

This technique has already been used by the writer in connection with a study of differences in speed of motor activity, as measured by the rate of movement during the solution of various performance tests (15). It was shown that Negro boys who had lived longer in New York reacted more quickly than those who had come more recently from the South. A somewhat similar method was used also by Peterson and Lanier in connection with intelligence tests (19); they divided their New York City group into "Northern born" and "born elsewhere," (that is, in the South and in the West Indies), and found a superiority in the former group. This superiority was not entirely reliable statistically, but it was slightly more than three times its probable error, which, considering the relatively small number of cases, seems reasonably significant.

The present investigation includes nine distinct studies made under the direction of the writer by candidates for the degree of Master of Arts in the Department of Psychology in Columbia University. Together they represent the findings on 3,081 subjects, consisting of ten and twelve-year-old Negro boys and girls in the Harlem schools; three of the studies were made with the National Intelligence Test, scale A, form I; three with the Stanford-Binet; one with the Otis Self-administering Examination, Intermediate Form; one with the Minnesota Paper Form

Boards; and one with an abbreviated Pintner-Paterson Performance Scale. The results of these studies will be presented separately, and also combined, wherever possible, so as to give a more general and at the same time more reliable picture of the environmental effect.

The three studies with the National Intelligence Test, scale A, form I, were made upon 1,697 twelve-year-old boys and girls in the Harlem schools in 1931 and 1932. In all three studies the subjects at the time of testing had passed their twelfth, and had not yet reached their thirteenth birthdays. The attempt was made in each case to secure every Negro boy or girl within this age range at the various schools at which the studies were made, and it is not likely that many were omitted. The scores were so combined as to make possible a comparison between a northern born control group and the southern born children who had been in New York one year, two years and so on up to eleven years. In every case note was taken of the average school grade of these various groups, so that degree of retardation or acceleration in school might also be used as a rough measure of present intellectual level. As might be expected, the intelligence-test scores and the school grades show a high degree of correspondence.

(1) The first of these studies was made by George Lapidus on 517 twelve-year-old boys between February and May, 1931; the subjects were all in attendance at three public schools and one junior high school in Harlem. The following table gives the average National Intelligence Test scores for each group.

TABLE 7

NATIONAL-INTELLIGENCE-TEST SCORE AND LENGTH OF
RESIDENCE (LAPIDUS)

Residence Years	1 Year	2 Years	3 Years	4 Years	5 Years	6 Years
No. of Cases	30	26	14	21	22	19
Average Score	64.43	63.96	54.50	75.09	76.72	67.21
Residence Years	7 Years	8 Years	9 Years	10 Years	11–12 Years	12 Years (Northern born)
No. of Cases	18	15	13	10	21	308
Average Score	86.61	79.93	74.00	79.10	93.85	86.93

These results are also presented in the following graph.

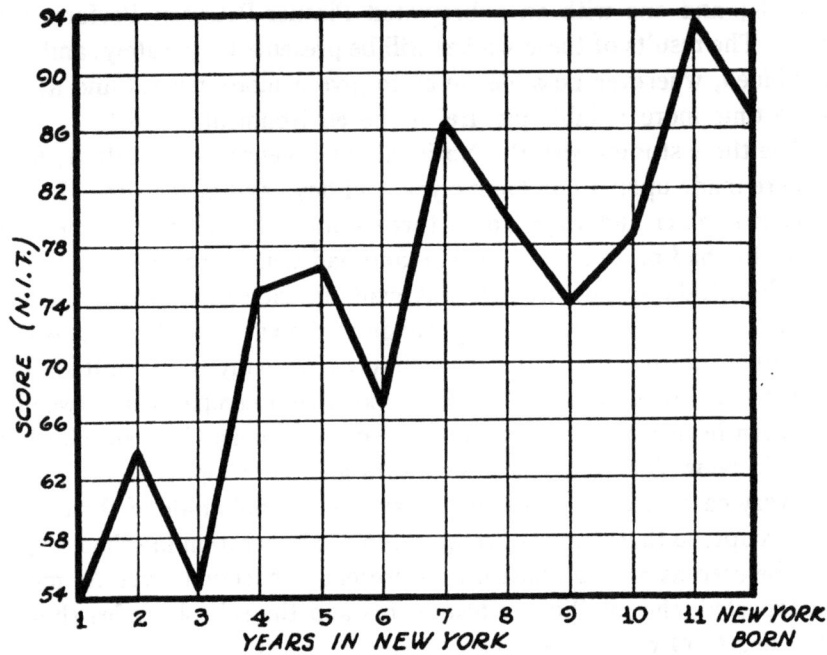

GRAPH 5. NATIONAL-INTELLIGENCE-TEST SCORES AND LENGTH OF
RESIDENCE IN NEW YORK

TWELVE-YEAR-OLD BOYS (LAPIDUS)

It may be seen that in spite of minor fluctuations there is a very
definite tendency for the scores to improve as length of residence increases. This result appears more clearly when the test scores of the
subjects are combined into two-year groupings; the rise is now definite
and regular.

TABLE 8

COMBINED GROUPS (LAPIDUS)

Residence Years	1 and 2 Years	3 and 4 Years	5 and 6 Years	7 and 8 Years	9, 10 and 11 Years	12 Years (Northern born)
No. of Cases	56	35	41	33	44	308
Average Score	64.21	66.86	72.32	83.58	84.64	86.93
Standard Deviation	29.5	29.5	33.2	29.1	30.4	28.9
Reliability of Av.	3.94	4.98	5.18	5.06	4.58	1.64

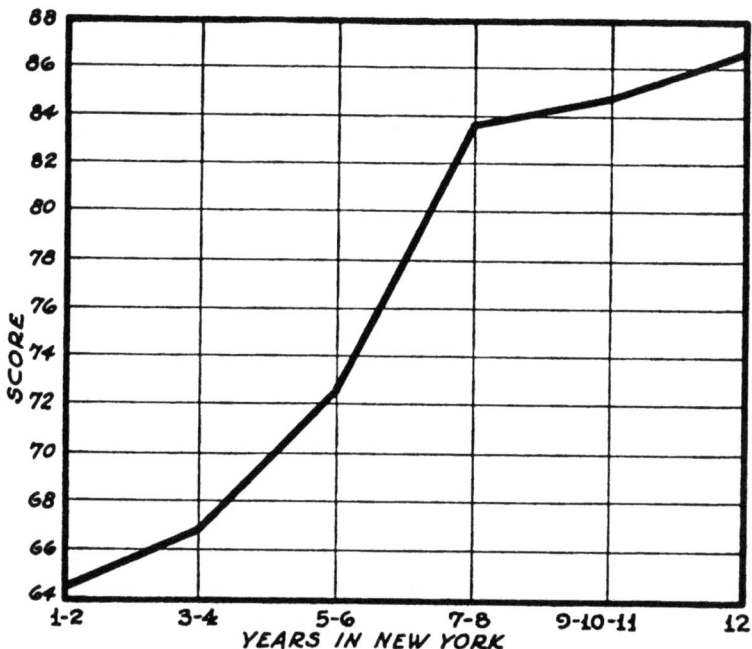

GRAPH 6. NATIONAL-INTELLIGENCE-TEST SCORES AND LENGTH OF
RESIDENCE

COMBINED GROUPS (LAPIDUS)

TABLE 9

GRADE RETARDATION (LAPIDUS)

Residence Years	1 Year	2 Years	3 Years	4 Years	5 Years	6 Years
No. of Cases	30	26	14	21	22	19
Average Grade	4B.86	4B.96	4B.69	5A.96	5B.38	5B.00
Standard Grade	7A for average 12-year-old pupils					
Retardation Years	2.07	2.02	2.15	1.52	1.31	1.50

Residence Years	7 Years	8 Years	9 Years	10 Years	11 Years	12 Years (Northern born)
No. of Cases	18	15	13	10	21	308
Average Grade	5B.84	6A.06	6A.25	6A.89	6A.04	6A.44
Standard Grade	7A	. . .	. . .	. . .	. . .	. . .
Retardation Years	1.08	0.97	0.88	0.55	0.98	0.78

Taking the northern born group as standard, there is a reliable difference in its favor over the one-two-year group and the three-four-year group, the difference divided by the sigma of the difference being equal to 5.33 and 3.83, respectively.* The superiority over the five-six-year group is practically reliable (99 chances in 100); over those subjects who have been in New York seven years or more the superiority is small and unreliable.

Table 9 gives the average grade for these various groups and also the degree of retardation. (Note: a grade of 4B.60 means that the average child in this group is 60 percent through 4B.)

It is clear in this case also that length of residence in New York has a very real effect upon scholastic level. Since 7A is the normal grade for twelve-year-old White pupils, it can be seen that the northern born group is only slightly more than three-fourths of a year retarded, and that the retardation is far more marked in the case of the recent arrivals from the South. This finding agrees with the experience of school authorities in the North, who have found it a very difficult problem to assimilate southern Negro children into their classes.

These results, which are in harmony with those reached in other investigations which form part of this study, correspond with the findings of Peterson and Lanier (19). The latter report that the New York Negro boys whom they tested were retarded approximately half a school year, whereas in Nashville the retardation is more than twice as great. There is a correlation of +0.49 between school grade and years of residence in New York for all subjects not born in New York City (72 cases); when those who had lived in New York more than six years were eliminated, the correlation was increased to +0.60 (40 subjects). There is no correlation between grade and length of residence in the city for those children who have been in school only in New York.

The authors comment:

All of these facts seem to point rather clearly to the conclusion that there are certain advantages of instruction and motivation in the educational system of New York which are superior to what the Negro gets in the South and in the West Indies. . . . Another point about the retardation of the transient Negro in New York should be mentioned. It is well known that any child transferring from one school system to another tends to be placed in a grade lower than

---

* When this ratio is greater then 3, the difference is conventionally regarded as a true or reliable one.

he would be in if he did not make the change. This may be due either to the poorer status of the first school or to differences in the curricula which make it necessary for the child either to go into the lower grade or to repeat a grade to avoid the omission of necessary training. As to the Negroes transferring from the South to New York, the teachers uniformly report that these children are very poorly trained. Any one who has given tests in Southern Negro schools knows that they do not have the facilities of the Whites, and so the comparative retardation of Southern Negroes in the New York schools is probably due largely to inadequate training, rather than to a more general tendency to place the newcomer back (19, pp. 17-18).

The investigation by the Chicago Race Commission (4) also showed that the great majority of retarded Negro children in Chicago were recent emigrants from the South; the northern born Negroes showed very little retardation. Reuter (23) points out further that late entrance into school life is the chief cause of the retardation of the Negro.

This in turn is due to the absence of school facilities, to over-crowded schools, to uneducated or indifferent parents, to poverty of the family, and to numerous other things that lie outside the personality of the child. The late entrance may or may not be related to low mental status. But granted a late entrance upon school life, whatever the cause, retardation through the grades and late graduation follow unless the late entrant be a child of more than average ability.

(2) The second study with the National Intelligence Test was made between February and May, 1932, by Charlotte Yates on 619 twelve-year-old girls in the Harlem schools. The scores were combined in the same way as in the preceding study, and the results are presented in Tables 10-14, and Graphs 7-9.

The results are again quite clear and definite (especially in the case of the groups combined in two-year intervals (Table 11 and Graph 8). The difference between the northern born group and the southern group with one to two years' residence in New York is completely reliable; d/sigma d (that is, the difference divided by the standard error of the difference) equals 5.87. For the three-four year group, d/sigma d equals 4.83. The difference in the case of all the other southern born groups is small and unreliable; in the case of the nine-ten-eleven-year groups, the difference is in favor of the southern born, but is also small and unreliable.

Table 12 gives the average grade and the degree of retardation of these various groups. The normal grade for twelve-year-old girls is 7A.

GRAPH 7. NATIONAL-INTELLIGENCE-TEST SCORES AND LENGTH OF
RESIDENCE IN NEW YORK

TWELVE-YEAR-OLD GIRLS (YATES)

TABLE 10

NATIONAL-INTELLIGENCE-TEST SCORE AND LENGTH OF
RESIDENCE (YATES)

Residence Years	1 Year	2 Years	3 Years	4 Years	5 Years	6 Years
No. of Cases	30	28	31	19	31	26
Average Score	63.66	70.43	76.25	88.78	96.58	94.38
Standard Deviation	25.3	34.96	18.78	26.23	24.79	21.24
Reliability of Average	4.61	6.60	3.55	6.02	4.40	4.16

Residence Years	7 Years	8 Years	9 Years	10 Years	11 Years	12 Years (Northern born)
No. of Cases	25	22	14	15	21	359
Average Score	96.56	98.09	106.85	94.67	99.23	97.86
Standard Deviation	20.46	25.88	18.10	29.76	27.95	29.7
Reliability of Average	4.26	5.51	4.83	7.69	6.1	1.5

It will be seen that there is again a very definite decrease in the amount of retardation, proportionate to length of residence in New York City. The retardation is appreciable only for those groups which

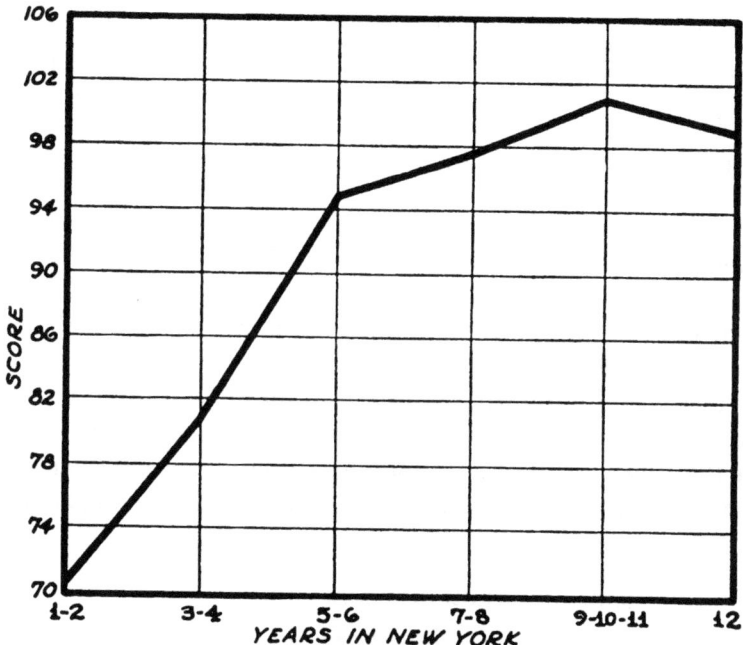

GRAPH 8. NATIONAL-INTELLIGENCE-TEST SCORES AND LENGTH OF
RESIDENCE
COMBINED GROUPS (YATES)

have been in New York six years or less; for all other groups it is only a small fraction of a year. There is practically no school retarda-

TABLE I I

COMBINED GROUPS (YATES)

Residence Years	1 and 2 Years	3 and 4 Years	5 and 6 Years	7 and 8 Years	9, 10 and 11 Years	12 Years
No. of Cases	58	50	57	45	50	359
Average Score	70.80	80.7	94.80	97.55	100	97.86
Standard Deviation	33.20	23	26.7	22.1	27.5	29.7
Reliability of Average	4.36	3.2	3.53	.32	3.87	1.5

TABLE 12

GRADE RETARDATION (YATES)

Residence Years	1 Year	2 Years	3 Years	4 Years	5 Years	6 Years
No. of Cases	30	28	31	19	31	26
Average Grade	4B.92	5B.76	5B.79	6A.09	6A.38	5B.94
Retardation Years	2.04	1.12	1.11	0.95	0.56	1.03

Residence Years	7 Years	8 Years	9 Years	10 Years	11 Years	12 Years
Number of Cases	23	22	14	14	21	359
Average Grade	6B.88	7A	6B.75	6B.75	6B.99	6B.84
Retardation Years	0.06	0.00	0.13	0.13	0.01	0.08

tion in the case of Negro girls who have had all of their schooling in New York City.

In this study a comparison was also made between the scores of those girls coming from *urban* and *rural* communities in the South. Since the school facilities in the southern cities are usually far superior to those in the country districts, it was felt that there might possibly

TABLE 13

MIGRANTS FROM CITY AND COUNTRY (YATES)

City Born Group

Residence Years	1 and 2 Years	3 and 4 Years	5 and 6 Years	7 and 8 Years	9, 10 and 11 Years
Number of Cases	47	37	33	37	36
Average Score	76	81.1	94.34	99.4	103.33
Standard Deviation	43.20	23.50	24.20	23.50	26.30
Reliability of Average	6.25	3.86	4.17	3.56	4.38
Average Grade	5B.89	5B.83	6A.45	6B.64	6B.88

Country Born Group

Residence Years	1 and 2 Years	3 and 4 Years	5 and 6 Years	7 and 8 Years	9, 10 and 11 Years
Number of Cases	9	9	11	7	4
Average Score	49.6	67.4	84	104	101.5
Standard Deviation	15.30	11.5	27.4	28.7	18.13
Reliability of Average	5.1	3.83	8.3	3.72	17.55
Average Grade	4B.70	5B.75	6A.15	6B.54	6B.88

be some difference between these two groups of migrants. Unfortunately, it was not always possible to determine very accurately the earlier residence of each child. In many cases the families had moved about a great deal before finally settling in New York. It frequently happened, for example, that a family moved from the country to the city in the South before coming North, and in that event the girl might give the name of the city as her previous residence. A study of this kind

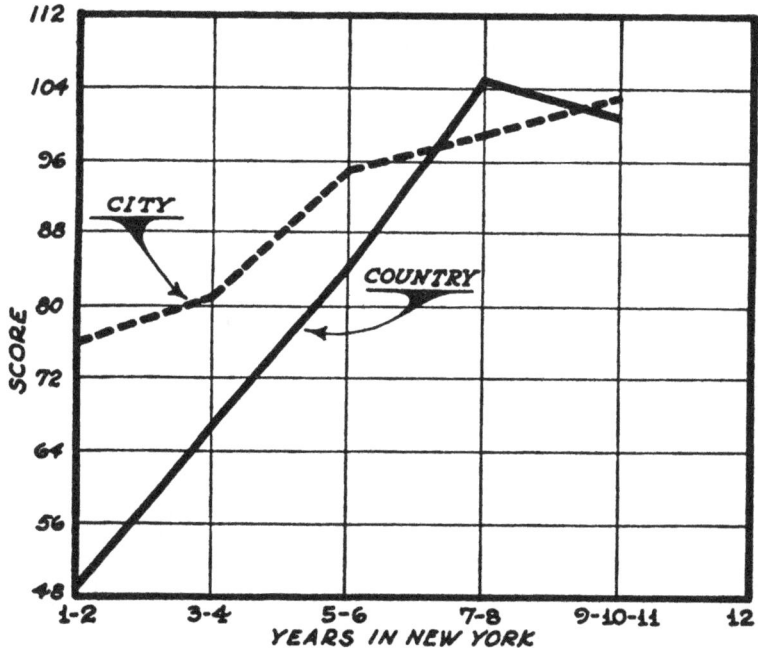

GRAPH 9. NATIONAL-INTELLIGENCE-TEST SCORES AND LENGTH OF RESIDENCE

CITY AND COUNTRY-BORN (YATES)

would require a much more careful personal inquiry into the movements of each family than was possible in this case.

Table 13 and Graph 9 present these results. The number of cases is slightly smaller than those reported in the other tables, as many children knew only the state and not the exact locality of their birth. The classification into city and country groups was based upon the census of 1927; a population of 5,000 inhabitants was regarded as constituting a city.

In spite of the small number of subjects from the rural districts, the results are very striking. They suggest that while the rural children start out far behind those from the city, after a number of years of residence in New York the difference disappears. In the case of the one-and-two-year groups, the difference in favor of the city born children is reliable; for the three-and-four-year groups it is almost reliable; for all the others it is small and unreliable.

Table 14 shows the degree of school retardation of city and country born children; there is again a marked difference between the city and country children who have been in New York only a short time, and no difference between the earlier arrivals.

TABLE 14

GRADE RETARDATION CITY BORN GROUP (YATES)

Residence Years	1 and 2 Years	3 and 4 Years	5 and 6 Years	7 and 8 Years	9, 10 and 11 Years
Number of Cases	47	37	33	37	36
Average Grade	5B.89	5B.83	6A.45	6B.64	6B.88
Retardation Years	1.05	1.09	0.53	0.18	0.06

GRADE RETARDATION COUNTRY BORN GROUP

Residence Years	1 and 2 Years	3 and 4 Years	5 and 6 Years	7 and 8 Years	9, 10 and 11 Years
Number of Cases	9	9	11	7	4
Average Grade	4B.70	5B.75	6A.15	6B.54	6B.88
Retardation Years	2.15	1.13	0.43	0.23	0.06

The comparison between city and country born children was repeated in the study by Marks (see below); unfortunately, the results were not nearly so definite.

(3) The third study with the National Intelligence Test was made between February and June, 1932, by Eli Marks on 561 twelve-year-old boys. Table 15 presents his results for the combined two-year groupings.

It will be seen that Marks' results are not nearly so definite as those reported in the other two studies. There is a general improvement among the southern groups, with the very marked exception of the one-and-two-year residence group, which is superior to all others except the northern born and those who have been in New York nine years or more. It is difficult to say just what factor has been responsible for this exception; it may be that the more recent arrivals represent a superior selection; it may be that schooling in the South has improved so markedly that the recent arrivals are better trained; it may also be that this is an accidental finding due to chance factors. Even in the case of the other groups the improvement dependent upon years of residence in New York is not by any means so clear as in the previ-

TABLE 15

NATIONAL-INTELLIGENCE-TEST SCORE AND LENGTH OF
RESIDENCE (MARKS)

Residence Years	1 and 2 Years	3 and 4 Years	5 and 6 Years	7 and 8 Years
No. of Cases	36	40	38	34
Average Score	87.53	78.70	81.18	85.82
Standard Deviation	29.9	37.5	28.4	31.5
Reliability of Average	5.0	5.9	4.6	5.4

Residence Years	9 Years and Over	Northern Born	Total Southern Group
No. of Cases	63	350	211
Average Score	96.19	90.78	87.02
Standard Deviation	32.3	35.1	32.9
Reliability of Average	4.1	1.9	2.3

ous studies. None of the differences between the standard (northern born) group and the various southern born combinations is reliable, though the chances are 97 in 100 of a true superiority over the three-and-four-year groups and the five-and-six-year groups. Those who have been in New York nine years or more are superior to the northern born, but the difference is not significant.

Table 16 gives the averages with the subjects from Panama and the West Indies omitted.

It will be seen that this omission does not alter the relative standing of the various groups to any considerable extent, although most of the averages are slightly lower as a result. It seems clear that the migrants from the West Indies rank somewhat higher than those from the South, probably because of the difference in previous school training, but the number of cases in these studies is too small to make any marked difference in the results.

Between those subjects born in New York City (212 cases) and the total northern born group (350 cases), there is no appreciable difference. The average score of the former is 91.34, of the latter, 90.78.

TABLE 16

RESULTS WITH WEST INDIAN AND PANAMA CASES OMITTED (MARKS)

Residence Years	1 and 2 Years	3 and 4 Years	5 and 6 Years	7 and 8 Years	9 Years and Over	Northern-born	Total Southern Group
No. of Cases	32	38	35	28	53	350	186
Average Score	85.59	80.63	80.17	84.75	92.94	90.78	85.153
Standard Deviation	29.7	37.2	28.8	33.3	31.3	35.1	32.6
Reliability of Average	5.3	6.0	4.9	6.3	4.3	1.9	2.4

Table 17 shows the average grade and the amount of retardation for the various combined groups.

There is a very definite improvement as length of residence increases. The one-two-year group, in spite of its excellent showing in the National-Intelligence-Test scores, was on the average more retarded than any other group. This suggests either a defect in the test, as far as correspondence with school grades is concerned, or inaccurate school placement of the newcomers in the New York schools. For the other groups

TABLE 17

AVERAGE GRADE AND GRADE RETARDATION (MARKS)

Residence Years	1 and 2 Years	3 and 4 Years	5 and 6 Years	7 and 8 Years	9 Years and Over	Northern Born
No. of Cases	35	36	37	32	61	340
Average Grade	5B.84	6A.17	6A.69	6B.06	7A.04	6B.82
Retardation Years	1.08	0.92	0.66	0.47	−0.02 (accelerated)	0.09

there is a close correspondence between intelligence rating and average grade. It will be seen that the northern born children, as well as those who have been in New York nine years or more, are not at all retarded. There is again evidence that a large part, if not all, of the retardation of Negro children in the New York City schools is due to the presence among them of a large number of newcomers from the South.

The study by Marks throws some light on an important problem which arises in connection with this whole investigation. If the subjects who have been in New York six years are superior to those who have been there only two, it is probable, as we have suggested, that length of residence in a superior environment definitely affects the test score. There is, however, another possibility. It may be that the quality of the more recent migrants is inferior to that of the earlier arrivals. The superiority of the six-year over the two-year group may be due, not to environmental influences, but to the fact that each year the northward migrants are inferior to those who preceded them. It is not very probable that such a difference would appear in successive

TABLE 18

COMPARISON OF 1931 AND 1932 AVERAGES

Group	1931 Average	1932 Average	Difference	Sigma Difference	Difference / Sigma Difference
1 and 2 Years	64.21	89.71	25.50	6.5	3.92
3 and 4 Years	66.86	79.06	12.20	7.8	1.56
5 and 6 Years	72.32	81.86	9.54	7.0	1.36
7 and 8 Years	83.58	85.06	1.48	7.6	.19
9 Years and Over	84.64	97.15	12.51	6.2	2.02
Northern Residence	86.93	90.23	3.30	2.5	1.32

years; one year or even two or three would hardly suffice to alter the conditions of migration sufficiently, although when migrants are compared after, let us say, a ten-year interval, such a difference in selective factors might possibly show an effect.

This problem arose in connection with the Army results as reported by Brigham (2). He pointed out that those European immigrants who had been in America longer scored higher in the Army tests than the more recent arrivals. His conclusion was that the migrants who came earliest were intellectually superior to those who followed. It may also be that those who have been longer in this country have had more time to learn the language and to acquire the information essential to high scores on the Army Alpha. (It should be added that Brigham no longer subscribes to the general position of his earlier writings (3).)

In the present investigation an attempt was made to throw light on this problem by having two studies made under exactly the same conditions, with similar subjects and the same test, but one year apart. The studies by Lapidus and Marks fulfilled these conditions; they were both on twelve-year-old boys with the National Intelligence Test, the study by Lapidus in 1931 and the one by Marks in 1932.

If the findings by Lapidus are due to a progressive deterioration in the quality of the migrants rather than to an environmental effect, the

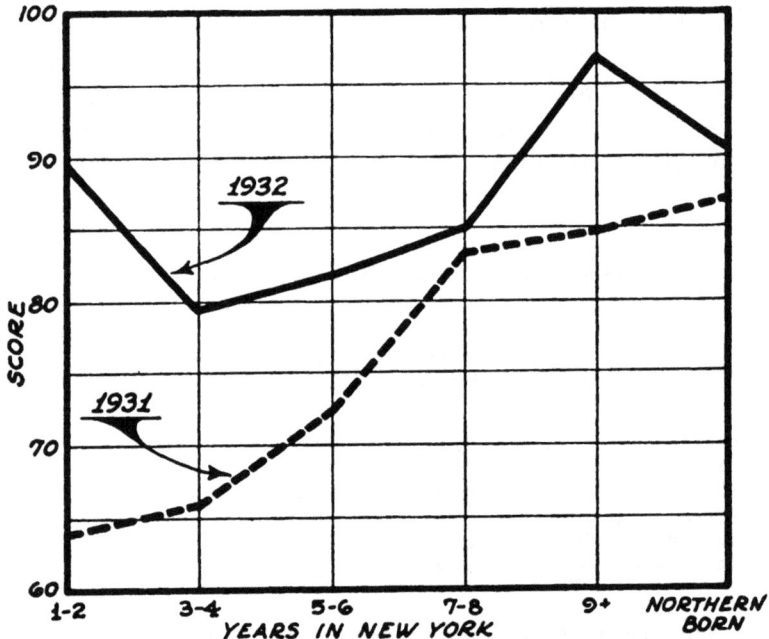

GRAPH 10. COMPARISON OF RESULTS OBTAINED IN 1931 AND 1932
(LAPIDUS AND MARKS)

results obtained by Marks in 1932 should be consistently *below* those obtained by Lapidus in 1931. A specific example will make this reasoning clearer. The twelve-year-old boys in the 1931 study who have been in New York four years, for example, arrived in 1927; those in the 1932 study who have been in New York for a similar period arrived in 1928. If the migrants are becoming inferior as time goes on, the four-year group in the later study ought to be inferior to the corresponding group in the earlier one. Table 18 and Graph 10 show the average scores obtained in the 1931 and the 1932 studies.

The 1932 averages are those of Table 12 with the omission of 90 subjects tested in schools which were not visited by Lapidus. Except for the one-and-two-year group, the difference between the 1931 and 1932 averages is not reliable. Whatever difference there is, however, is consistently in favor of the 1932 group, that is, of the more *recent* arrivals. This difference may be due to improvement in the schooling in the South; in any case there is no evidence that the more recent

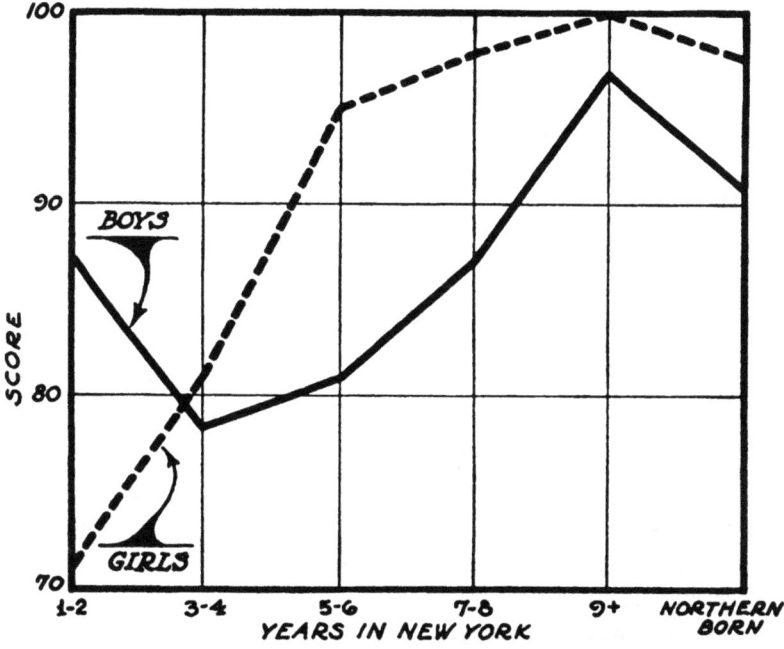

GRAPH 11. COMPARISON OF BOYS AND GIRLS
(MARKS AND YATES)

arrivals are inferior. The conclusion is therefore justified that the superior showing of those subjects who have had a longer period of residence, is due to this longer residence, and not to any regular change in the quality of the migrants.

This conclusion is strengthened by the results reported in the study by Hand (see above). The comparison of the school records made by migrants in the various years from 1915 to 1930 showed on the whole a tendency toward improvement in more recent years. If anything, the recent migrants are better, not poorer, than the earlier ones, and it is impossible to assume that a less rigid selection is now bringing a less intelligent migrant North.

One final comparison was made possible by Marks' study. Yates' study of girls and Marks' study of boys were made at about the same time and under approximately the same conditions. Table 19 and Graph 11 show the average scores obtained by the two groups.

TABLE 19

COMPARISON OF BOYS AND GIRLS

Group	Girls Average	Boys Average	Difference	Sigma Difference	Difference / Sigma Difference
1 and 2 Years	70.80	87.53	−16.73	6.6	−2.53
3 and 4 Years	80.70	78.70	2.00	6.7	.30
5 and 6 Years	94.80	81.18	13.62	5.8	2.35
7 and 8 Years	97.55	85.82	11.73	6.3	1.86
9 Years and Over	100.00	96.19	3.81	5.6	.68
Northern Residence	97.86	90.78	7.08	2.4	2.95

None of the differences is reliable, but a number of them are almost reliable and, with the exception of the one-and-two-years' residence groups, all are in favor of the girls. The National Intelligence Test is one in which linguistic abilities play a relatively important part, and this finding is therefore consistent with the demonstration that among White children girls do somewhat better than boys on tests in which linguistic factors predominate.

(4) As the three studies with the National Intelligence Test were made under the same conditions, the results were combined to show more clearly the extent of the environmental effect.

TABLE 20

THE THREE STUDIES COMBINED (LAPIDUS, YATES, AND MARKS)

Residence Years	1 and 2 Years	3 and 4 Years	5 and 6 Years	7 and 8 Years	9 Years and Over	Northern Born
No. of Cases	150	125	136	112	157	1017
Average Score	72	76	84	90	94	92

The improvement with length of residence is clear and definite. The excellent showing of the one-two-year group in Marks' study raises the

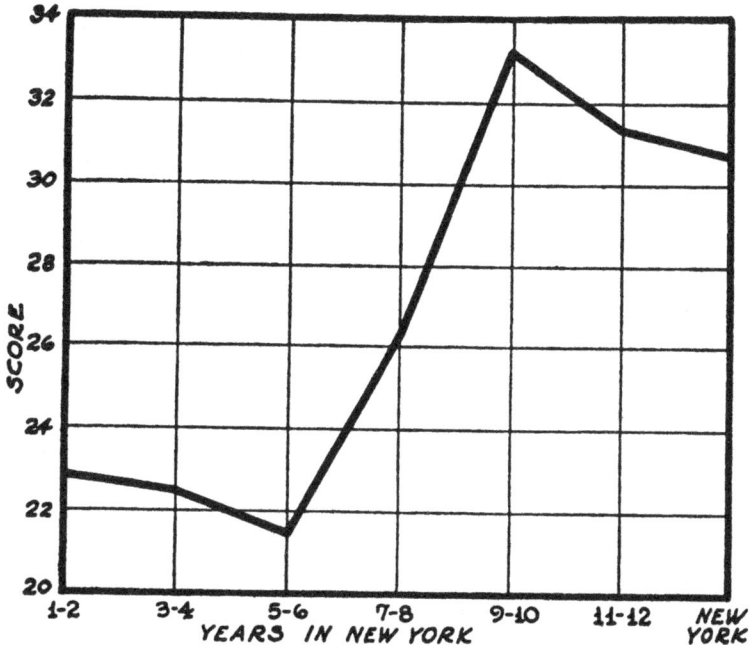

GRAPH 12. OTIS SCORES AND LENGTH OF RESIDENCE
TWELVE-YEAR-OLD GIRLS (TRAVER)

level of that group considerably, but not above that of any of the succeeding year combinations. It will be noticed that the range of average
scores is from 74 for the one-two-year group to 92 for the control group;
this suggests that the I.Q. remains constant only when there is relative
constancy in the environment.

TABLE 21

OTIS SCORES AND LENGTH OF RESIDENCE (TRAVER)

Residence Years	1 and 2 Years	3 and 4 Years	5 and 6 Years	7 and 8 Years	9 and 10 Years
No. of Cases	28	37	45	19	23
Average Score	22.8	22.5	21.5	26.2	33.1
Standard Deviation	12.6	9.6	10.6	13.7	13.7

Residence Years	11 Years and Over	New York Born	Total Southern Group
No. of Cases	18	243	170
Average Score	31.4	30.9	25.1
Standard Deviation	13.8	15.1	13.0

(5) One study was made of 536 twelve-year-old girls, between February and May, 1931, by Isabel D. Traver, with the Otis self-administering examination, Intermediate form.

These results are not nearly so definite as those obtained in the National Intelligence Test studies. There is practically no difference between the one-two, three-four and five-six-year groups; it is only with the seven-eight-year group that any correspondence between test score and length of residence in New York becomes evident. Whether this is a function of the test or of accidents of sampling, it is impossible to say. In any case there is still a very marked difference between the earlier arrivals (six years or less) and the later ones (seven years or more), and this is clearly in favor of the latter. While the environmental effect does not appear very early in this study, it is still there. The northern born group is reliably superior to the one-two, three-four, and five-six-year groups; it is definitely, but not quite reliably, superior to the seven-eight-year group, and only slightly and unreliably superior to those of more than eight years' residence in New York. The northern group as a whole is also reliably superior to the southern born group as a whole. There was no difference between the girls born in the West Indies (30 cases) and those born elsewhere in the South (170), and only a slight superiority of those born in New York City over those born elsewhere in the North. In general it may be said that these results are corroborative of those found in the National Intelligence Test studies.

CHAPTER V

NEW YORK CITY—STANFORD-BINET TESTS

Three studies were made by means of the Stanford-Binet test, two of them being of ten-year-old girls, and one of ten-year-old boys. All the testing in these studies was of course individual, and the number of subjects was consequently much smaller than in the case of the group tests. The results are even clearer than in the studies reported above, and the evidence for an environmental effect is unmistakable. This is especially interesting in view of the frequently expressed opinion that the Binet, unlike the linguistic group tests, is relatively free from the influence of nurture and background, and is a more direct test of native ability.

(1) The first of these studies was made by Jeannette Skladman; the subjects were 107 ten-year-old girls in the Harlem public schools, and, as in the other studies, they were divided on the basis of years of residence in New York.

TABLE 22

STANFORD-BINET INTELLIGENCE QUOTIENTS AND LENGTH OF NEW YORK
RESIDENCE, TEN-YEAR-OLD GIRLS (SKLADMAN)

Group	N	Average I.Q.	S.D.
Less than one Year	20	81.8	9.14
1-2 Years	20	85.8	7.91
3-4 Years	19	90.3	8.42
More than 4 Years	20	94.1	12.6
New York born	28	98.5	9.47

The results show a surprisingly large difference between the New York born and the recent arrivals. The former group is reliably superior to all the groups with less than four years' residence. There is also a reliable difference between the southern born with more than four years' residence in New York and those with less than one year's residence. The New York born group has an intelligence quotient of 98.5, that is, almost exactly at the norm for white children. If the results of this study can be accepted as they stand, they suggest that the New York environment is capable of raising the intellectual level of the Negro children to a point equal to that of the Whites. The number of cases is small, however, and the results are not entirely consistent with those obtained by other investigators, nor with the other studies which form a part of the present investigation. The only other studies in the knowledge of the writer which showed the Negroes to be the equals of the Whites is that by Peterson and Lanier (19) referred to above, and that by Graham (10) in Georgia. Practically all other investigations agree in placing the northern Negro definitely above the Negro from the South, but still somewhat below the northern White.

The results also show clearly the improvement in school grade with increased length of residence, as well as the fact that the children born in New York City show hardly any degree of retardation.

Miss Skladman points out that the chief deficiency of the New York born group was in the vocabulary tests. Only one child passed the twelve-year vocabulary requirement, while most of them did not reach the ten-year level, and a few failed to reach even the eight-year level. Their vocabulary scores were still, however, markedly superior to those of the more recent arrivals.

(2) A second study of ten-year-old girls by means of the Binet test was made by Elsie Wallach between February and May, 1932. In this study there were 167 subjects, 118 of these southern born. All the girls who had lived in New York less than four years were tested; the other subjects, namely those born in New York, or with more than four years' residence, were chosen at random from a list arranged in alphabetical order. The results follow.

TABLE 23

STANFORD-BINET INTELLIGENCE QUOTIENTS AND LENGTH OF NEW YORK
RESIDENCE, TEN-YEAR-OLD GIRLS (WALLACH)

Group	N	Average I.Q.
Less than one Year	24	80.5
1–2 Years	23	84.0
2–3 Years	21	85.9
3–4 Years	24	85.1
More than 4 Years	26	87.1
New York born	49	85.2
All Southern born	118	84.5

The results again indicate an improvement in the I.Q. almost directly proportionate to length of residence, but in this case the improvement is not nearly so marked, and the New York City control group has an I.Q. of 89.3, that is, still definitely below the White norm. The difference between the best and the poorest (less-than-one-year) groups is about 6 I.Q. points; this difference is almost reliable (99 chances in 100). There is a similar difference (99 chances in 100) between the more-than-four-years group and the less-than-one-year group; the chances are 98 in 100 that the two-three-year group is also superior to the less-than-one-year group. While none of these differences is completely reliable statistically, the general trend is unmistakable, even though the improvement with length of residence is not nearly so marked as in the case of Skladman's study.

(3) Henry Rogosin gave the Binet test to 147 ten-year-old boys. His study paralleled in every way that by Wallach, and was carried on at the same time and under exactly the same conditions. The results follow.

TABLE 24

STANFORD-BINET INTELLIGENCE QUOTIENTS AND LENGTH OF NEW YORK
RESIDENCE, TEN-YEAR-OLD BOYS (ROGOSIN)

Group	N	Average I.Q.
Less than One Year	18	82.6
1–2 Years	17	84.5
2–3 Years	19	83.0
3–4 Years	22	85.9
More than 4 Years	21	87.7
New York born	50	89.3
All Southern born	97	84.9

These results are in general quite similar to those obtained by Miss Wallach, except that the improvement is less definite among the groups with shorter residence in New York, and somewhat more definite in the case of those with longer residence. The New York born group has an I.Q. of 89.3, again about 6½ points above the one-year group, but still definitely below the White norms. In this case the sex difference is in favor of the boys: they are superior throughout the study, although not reliably so.

The improvement with increased length of residence is again unmistakable, although the superiority of the New York control group over the others is not reliable. There are, however, 98 chances in 100 that the New York group is superior to the less-than-one-year group, 95 chances in 100 that it is superior to the one-two-year group, 97 chances in 100 that it is superior to the two-three-year group, and 99 chances in 100 that it is superior to the whole southern born group combined.

There is a similar relationship between school grade and years of residence.

(4) The results of these last two studies were combined; they are presented in Table 25.

TABLE 25

STANFORD-BINET INTELLIGENCE QUOTIENTS AND LENGTH OF NEW YORK RESIDENCE, TEN-YEAR-OLD BOYS AND GIRLS COMBINED

Group	N	Average I.Q.
Less than One Year	42	81.4
1–2 Years	40	84.2
2–3 Years	40	84.5
3–4 Years	46	88.5
More than 4 Years	47	87.4
New York born	99	87.3
Total Southern born	215	84.7

When the groups are combined in this way there is a reliable superiority of the New York born group over the less-than-one-year group, and a reliable superiority also of the more-than-four-years group over the less-than-one-year group. None of the other differences is completely reliable, although many of them approach reliability very closely. There is again a difference of about 6 I.Q. points between the New York born group and the less-than-one-year group, and although the differences among adjacent groups are small, the general trend is again unmistakable.

CHAPTER VI

NEW YORK CITY—PERFORMANCE TESTS

Two studies were made by means of performance tests. In the first of these B. H. Brown gave six tests in the Pintner-Paterson series to 110 ten-year-old boys. The tests were the triangle test, the Healy "A," the two-figure board, the five-figure board, the casuist form board, and the Knox cube test. All the tests were administered and scored according to the directions given in the Pintner-Paterson *Manual* (22). The total score for each child was computed both according to the method of median mental age and the method of total points on the point scale. The southern born group was divided into three subgroups: (1) those who had lived less than two years in New York; (2) more than two years and less than five years; and (3) more than five years. The results are presented in the following table and graph.

TABLE 26

PINTNER-PATERSON SCORES AND LENGTH OF RESIDENCE (BROWN)

Group	N	Av. Mental Age	S.D.	Total Points, Av.	S.D.
Less than 2 Years	20	7.25	3.03	142.5	53.5
2-5 Years	20	7.65	1.85	139.8	48.4
More than 5 Years	20	7.50	2.29	152.1	46.6
Total southern	60	7.47	2.44	144.8	49.8
Northern born	50	8.65	2.17	164.5	50.2

The correspondence between mental level and years of residence is not so definite as in the studies previously reported, but there are still fairly clear evidences of an environmental effect. The average mental age (M.A.) of the less-than-two-years group is the lowest, and that of the northern born group, the highest; the difference between the two is almost statistically reliable (94 chances in 100). The other two southern groups have a slightly higher M.A. than the very recent arrivals, and are definitely below the New York born. There is an almost completely reliable superiority of the northern born group over the combined southern groups (99 chances in 100). With the point scale the results tend in the same direction. The two-to-five-years group is, however, slightly lower than the less-than-two-years group; the more-than-five-years group is higher than either, and the northern born group is again the highest. There are 98 chances in 100 that the northern born group is superior to all the southern groups combined.

In the light of an earlier investigation by the writer which dealt with the question of environmental influences on speed of movement (15),

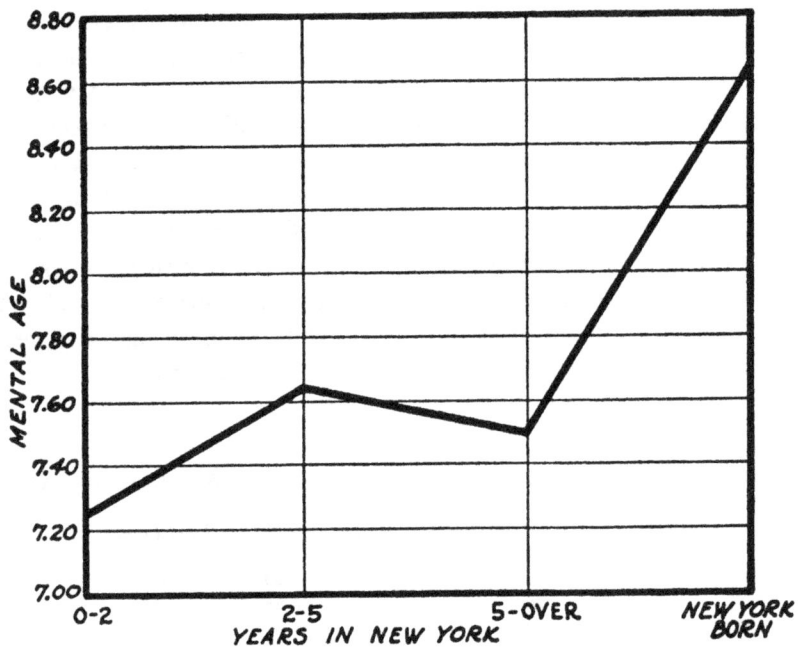

GRAPH 13. PINTNER-PATERSON MENTAL AGE AND LENGTH OF RESIDENCE IN NEW YORK (BROWN)

a speed index was calculated for each of the subjects in the present study. This was done by recording the number of moves per minute during the process of solving Healy "A." The results are presented in Table 27. It will be seen that there are very small and insignificant differences between the various groups, and that there is no observable relation between speed of movement in this test and length of residence in New York City. This finding is in conflict with that reported by the writer in the study referred to above, in which there appeared to be a definite increase in speed proportionate to length of residence. Further studies are required to settle this point.

In this connection a dotting test was also administered to these subjects; the test was simply to put a dot in each square on a piece of

TABLE 27

SPEED OF MOVEMENT, HEALY "A" (BROWN)

Group	N	Moves per Minute	S.D.
Less than 2 Years	20	21.6	4.8
2–5 Years	20	20.2	4.9
More than 5 Years	20	21.6	4.4
Total southern	60	21.1	4.7
Northern born	50	21.1	6.2

squared paper as quickly as possible. The score is the total number of squares dotted in one minute. The results are presented in Table 28. In this case, although the differences are small, the correspondence between length of residence and speed seems to be in the opposite direction. The northern born group is the slowest, and the two-to-five years group the fastest. All these differences are, however, small and unreliable.

The second study with performance tests was made by Eugene Horowitz, who gave the Minnesota paper-form-board (see 18) to 416 twelve-year-old boys. The tests were administered according to the directions in the *Manual*, except that the method of scoring used in this study was not quite the same. As we have no special interest in comparing the results of this study with the norms established by the

TABLE 28

SPEED OF MOVEMENT, DOTTING TEST (BROWN)

Group	N	Dots per Minute	S.D.
Less than 2 Years	20	97.7	22.4
2–5 Years	20	102.0	13.8
More than 5 Years	20	98.6	13.8
Total southern	60	99.4	17.2
Northern born	50	96.3	17.9

Minnesota group, and as the same method of scoring was of course used throughout the present study, this deviation from the usual procedure will hardly affect the results or the conclusions. The reason for the change was that the Minnesota method of scoring was found to be

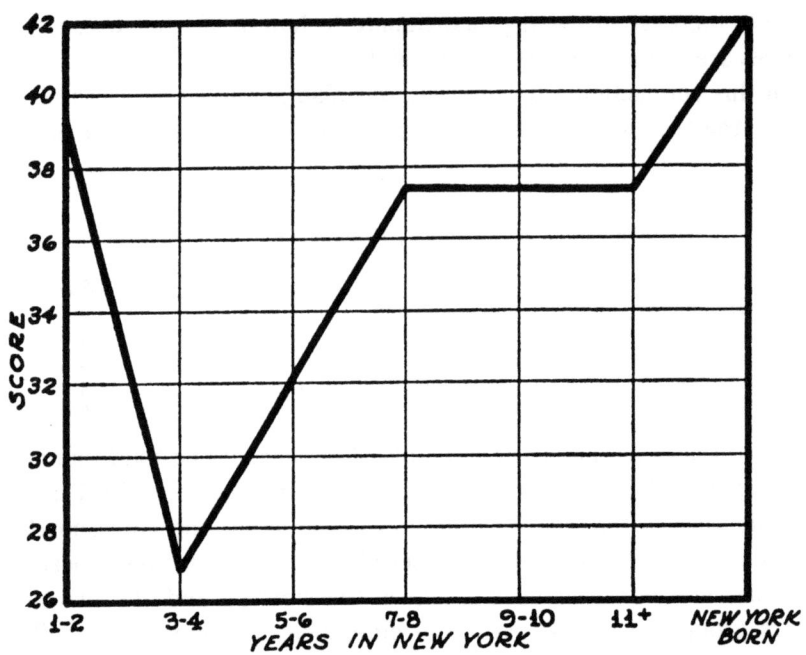

GRAPH 14. MINNESOTA-FORM-BOARD SCORES AND LENGTH OF
RESIDENCE (HOROWITZ)

too drastic for the younger subjects used in this study. The results are
presented in Table 29 and Graph 14.

TABLE 29

PAPER-FORM-BOARD SCORES AND LENGTH OF RESIDENCE

Group	Residence Years	Median	Q (Semi-interquartile Range)	Number of Cases
A	1–2	39.00	30.00	27
B	3–4	26.67	33.34	25
C	5–6	31.88	24.14	30
D	7–8	37.50	17.79	23
E	9–10	37.50	24.25	25
F	11–	37.50	38.44	41
G	Native born, New York City	41.61	26.73	223
H	Northern born	42.63	26.83	265
I	Southern born	35.22	24.52	151

These results seem at first glance to show no relation whatever between
score and length of residence. What seems to have happened, however,

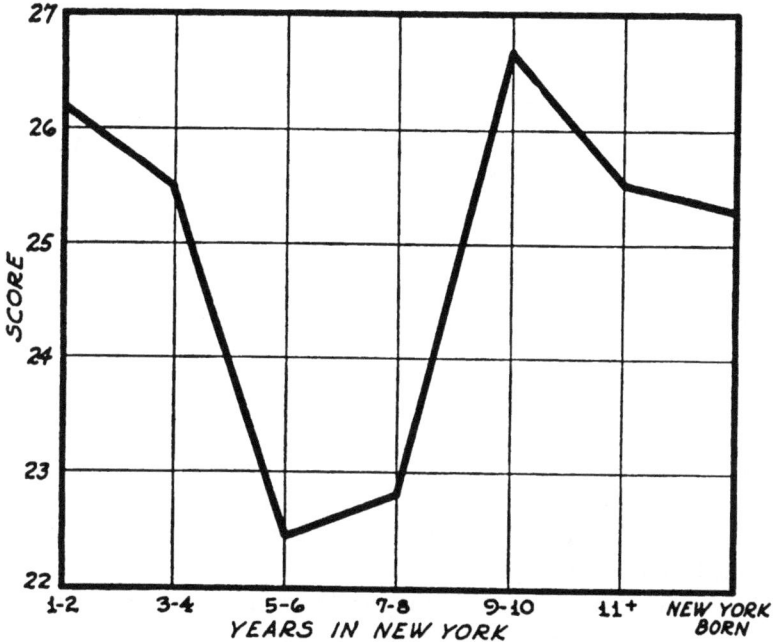

GRAPH 15. CURTIS ARITHMETIC SCORES AND LENGTH OF
RESIDENCE (HOROWITZ)

is that a particularly good showing was made by the very recent arrivals (less-than-two-years group); with the exception of this one group
the other results do show such a correspondence very definitely. The
subjects in this study were the same as those used by Eli Marks
and in his study it was also the very recent arrivals whose records in-
terfered with this correspondence. Evidently this group is a particularly
good one; just why this should be is not apparent.

All the differences, however, are unreliable. The largest difference,
that between the northern born group and the three-to-five-years group
has only 88 chances in 100 of being a true difference. As far as the
study as a whole is concerned, the only safe conclusion appears to be
that there is no proof of any very definite correspondence between
length of residence and the scores on this type of group performance
test. There are, however, 94 chances in 100 that the northern born group
is superior to the combined southern born groups.

Horowitz also administered to these same subjects the Curtis arithmetic test. The results are presented in Table 30 and Graph 15.

TABLE 30

CURTIS ARITHMETIC TEST SCORES AND LENGTH OF RESIDENCE (HOROWITZ)

Group	Residence Years	Median	Q (Semi-interquartile Range)	Number of Cases
A	1-2	26.25	4.52	30
B	3-4	25.50	6.78	34
C	5-6	22.50	6.37	38
D	7-8	22.86	7.00	29
E	9-10	26.67	6.88	33
F	11-	25.50	5.00	22
G	Native born, New York City	25.30	5.39	245
H	Northern born	25.69	5.04	299
I	Southern born	24.67	6.39	186

These results require little comment. There seems to be no evidence of any relation between test score and length of residence. There are only 84 chances in 100 that the northern born group is superior to the combined southern groups.

The studies reported in this chapter have failed to demonstrate any definite environmental effect. It is true that there are indications of such an effect in the two form-board studies, but neither of these is sufficiently clear cut to permit of a dependable conclusion. When taken in conjunction with the results reported in preceding chapters, they justify the inference that those tests in which school training enters are very much more affected by a change of environment than those in which the effect of schooling is reduced to a minimum. The environment probably has some effect on these latter also, but it is not nearly so marked nor so immediate. One is tempted also to draw the conclusion that the results with performance tests show that there is merely a "linguistic," and not a *real* intellectual difference between the northern and the southern born groups, and that the differences which appear are entirely a function of the environmental effect on the particular tests used. The writer inclines to that view, but does not feel that the material presented here justifies it entirely, since the exact relation between the abilities measured by linguistic tests and those measured by performance tests is still in doubt.

CHAPTER VII

CITY AND COUNTRY

The comparison of city and country children by means of intelligence tests involves problems very similar to those we have discussed in the preceding chapters. The large number of investigations of the intelligence of children in rural communities has almost without exception shown them to be definitely inferior. There is the same difficulty in deciding whether this inferiority is due to the less favorable environmental conditions in the country, or to the fact that selective migration has drained off from the smaller communities their most capable and intelligent families.

Pintner (21) after a survey of these results concludes that they cannot be entirely accounted for by environmental differences, and that selective migration has undoubtedly played a part. He does not, however, advance any very direct evidence in favor of this view. Shimberg (24) also summarizes the results and finds the same general trend in favor of the city children, but inclines to the opinion that it is unfair to compare city and country children with tests standardized only on the former; such tests may be applicable to city children but should not be applied without modification to children living in a totally different environment. She was able to demonstrate that an information test standardized on rural children placed children from the city at an equal disadvantage, and therefore showed them to be somewhat "inferior."

The present writer's earlier study (16) on various racial and national groups in Europe demonstrated the consistent superiority of the children in three large European cities, Paris, Hamburg, and Rome, over all the rural groups tested in France, Germany, and Italy. This study was made by means of a number of tests in the Pintner-Paterson performance series, and it is probable that differences in speed played an important part in the results. That study also raised but did not attempt to answer the question of whether or not there is selective migration from country to city.

In the present study of Negro children the attempt was also made to see whether the city environment had any effect in raising the test scores of children born in the country. It is obvious that there is a very marked difference in the educational and cultural backgrounds of rural and urban Negro communities in the South; it is even probable that in many cases the difference in opportunity is greater than that between North and South. In this part of the study it was not found possible to use early school records made by the migrants before they reached the city, as the rural schools rarely kept record books which might be considered adequate. There were, however, in the three southern cities visited—New Orleans, Nashville, and Atlanta—a large number of children in the Negro public schools who had come from surrounding rural communities and who differed in the number of years of their residence in the city. In their case also, therefore, the same

TABLE 31

TEST SCORE AND LENGTH OF RESIDENCE IN CITY

Years	Number of Cases	Average Score
1	39	38.3
2	25	43.2
3	36	44.7
4	47	62.5
5	52	56.2
6	53	62.2
7 or more	165	68.7
City born	359	74.6

technique was applied as in the New York studies, and the results were analyzed to see whether there was any relation between test scores and length of exposure to a more favorable environment.

The National Intelligence Test, Scale A, Form I, was administered by the writer in 1930 to 786 twelve-year-old Negro boys in the public schools of New Orleans, Atlanta, and Nashville. The results are shown in Table 31. It will be noticed that in this study those boys who had been in the city seven years or more were included in one group; it was felt that this could legitimately be done since all of them had had all of their schooling in one or another of the large cities.

In addition a small group of boys from the little village of Thibodaux, La., was also tested. There were only 11 subjects and the results therefore cannot be taken too seriously, but it is at least suggestive that their score was 47, or slightly above that obtained by the group of migrants who had lived less than three years in the city. A more extensive study of rural children is indicated, but as far as these results go, they suggest that the migrants to the city are not superior at the outset to those who remained in the country, but that their later superiority is due to the gradual influence of the better environment. The improvement is rapid and definite; with the one exception of the four-year group, the relationship between average score and length of residence is perfect. There is a statistically reliable superiority of the city born group over all the country born groups, as well as a statistically reliable superiority of all the groups which have been in the city four years or more over those which have been there three years or less. There can be no doubt in this case that the environment plays an exceedingly important part in determining the test score.

CHAPTER VIII

SCHOOLS NORTH AND SOUTH

To anyone with any acquaintance with the Negro school system in the South, it will hardly come as a surprise that a period of residence in the superior New York school environment should improve the intelligence of Negro children. This is not the occasion for a discussion of the intricate relationship between schooling and intelligence as measured by the intelligence test; in the ordinary linguistic tests (of the type of the Binet or the National Intelligence Test) the success in many of the tasks to be performed depends so clearly upon the type of training received in school that it is not surprising that better schooling should mean a higher intelligence quotient. The opinion expressed in the earlier days of the testing movement in America, that the tests measure native endowment altogether apart from the influence of training and background, is now held by few if any of the psychologists who have concerned themselves with tests. The question is no longer whether training has an effect, but rather how great that effect can be.

Better schooling is of course not the only environmental factor which influences the test scores, but it is probably the most important one. The other factors—superior economic status, better opportunities for "extra-curricular" activities, greater motivation resulting from a better chance of success—also play their part, although it is a part which cannot very easily be evaluated. The difference in the school systems in the North and the South is, however, clear and definite, and the influence of the change upon the southern Negro child, especially when that change occurs early in life, can hardly be overestimated.

Woofter (27) comments as follows on the southern schools:

The South is not only poorer than the North, but also less disposed to distribute such funds as are available according to the school population. The Negro schools are a secondary consideration. In comparison with schools for White children they have fewer seats in proportion to the school population, more pupils per teacher, more double sessions, fewer teachers, poorer salaries, fewer and smaller playgrounds, and less adequate provision for the health and comfort of pupils and teachers.

The per-capita expenditure for the education of White and Negro children in the South shows an illuminating contrast. The report of the Phelps-Stokes fund for 1910-1920 (20) states that

the per capita in the Southern states was found to be $10.32 for each White child, and $2.89 for each Colored child. The per capita figures for the different states vary widely. In the border states where the proportion of Negroes is relatively small, the per capita for Negroes is higher than in the other states. The most striking differences, however, are in the county expenditures. State school funds are apportioned to each county on the basis of population without regard to race. The county officers then divide these funds according to their interpretation of the needs of the White and Colored pupils.

The report gives a table of expenditures in southern counties, arranged according to the proportion of Negroes in each county.

TABLE 32

PER CAPITA EXPENDITURE ON EDUCATION

County Groups	White	Negro
Under 10 percent Negro	$7.96	$7.23
10–25 percent Negro	9.55	5.55
25–50 percent Negro	11.11	3.19
50–75 percent Negro	12.53	1.77
75 percent and over Negro	22.22	1.78

Charles S. Johnson also points out the wide disparity in per-capita expenditures for White and Negro children in the southern states (13). For example, South Carolina spends $4.48 per Negro child, and $45.45 per White child; the figures for Alabama are $5.45 and $37.63; for Georgia, $7.44 and $35.24; for Louisiana, $8.02 and $46.67; for Mississippi, $9.34 and $42.17; for Virginia, $14.86 and $54.21. In Tennessee the proportion is considerably more equitable, $20.15 to $33.47. In the border states, Kentucky and West Virginia, the per-capita expenditures for the two groups are about the same.

Embree in *Brown America* (5) comments:

Studies of eight Southern states show average expenditures of $44.31 per capita for Whites and only $12.50 for Negroes. . . . The inadequacy of these provisions for either race is seen when one compares them with the average expenditure throughout the United States as a whole, which is $87.22 per school child.

In the light of these figures the educational handicap under which the average southern Negro child suffers hardly requires further comment.

It may be noted in this connection that Clark Foreman (7), who administered achievement tests to colored children in a number of rural counties in Georgia and Alabama, found a close correspondence between the standing of the various counties in these tests and the percapita expenditure for the education of Negro children.

The marked discrepancy between educational facilities in the North and in the South throws considerable light on the question of school retardation among Negro children. A large number of the children who are retarded in the northern schools have come from the South and have suffered certain educational handicaps which they have not yet been able to overcome. The data on school retardation presented elsewhere in this study show that northern born children are only very slightly over age, and that the degree of retardation is very much greater among the newcomers from the South. Similar results are reported by Johnson (13) for the Negro school population of Detroit. In the Detroit schools Negro children born in Michigan show only 4.76 percent retardation, while those born in the South are very much more retarded, in proportions closely approximating the condition of the school system in their home states; for example, children from Virginia show 20 percent retardation, and those from Mississippi, 25 percent. Another important factor in retardation is delay in entering school, rather than slow progress. Woofter (27) points out that a great many of the colored children are "merely pedagogically retarded, not mentally deficient, and these tend to progress faster and to catch up with their normal grade."

The picture is a clear one. The southern states are relatively poor, have much less money to expend on elementary and secondary education, and divide that money in such a way that Negro children obtain far less than their proportionate share of educational opportunities. This discrepancy is more marked in some states than in others, and more marked also in rural than in urban districts, but in general it is certainly fair to assume that those Negro children who have received part of their education in the South are placed at a definite disadvantage in competition with others. Those who have come North at an early age, and have had all or nearly all their schooling in the North, have been very nearly able to overcome the handicap; those who have come North in recent years still suffer from it very definitely.

CHAPTER IX

DISCUSSION

There seems to the writer to be no reasonable doubt as to the conclusion of this study. As far as the results go, they show quite definitely
that the superiority of the northern over the southern Negroes, and the
tendency of northern Negroes to approximate the scores of the Whites,
are due to factors in the environment, and not to selective migration.

There is, in fact, no evidence whatever in favor of selective migration. The school records of those who migrated did not demonstrate any
superiority over those who remained behind. The intelligence tests
showed no superiority of recent arrivals in the North over those of the
same age and sex who were still in the southern cities. There is, on
the other hand, very definite evidence that an improved environment,
whether it be the southern city as contrasted with the neighboring
rural districts, or the northern city as contrasted with the South as a
whole, raises the test scores considerably; this rise in "intelligence" is
roughly proportionate to length of residence in the more favorable environment.

Even under these better environmental conditions Negro children
do not on the average quite reach the White norms. Since the environment of the New York Negro child is by no means the same as that
of the White, except perhaps as far as schooling is concerned, this
result does not prove that the Negro is incapable of reaching the White
level. As has been pointed out elsewhere, Negro communities even in
the North represent more or less isolated self-sufficient groups; their
background and their ways of life are not at all the same as those of
the larger group of which in other respects they form a part. While we
have no complete proof that an improvement in their background can
bring them up to the White level, we also have no right to conclude the
opposite. What we can safely say is that as the background improves,
so do the scores of the Negroes approximate more and more closely the
standards set by the Whites. The final and crucial comparison could
only be made in a society in which the Negro lived on terms of complete

equality with the White, and where he suffered not the slightest social, economic, or educational handicap. It is doubtful whether such a society exists, but perhaps some approximation to it could be found in, for example, Brazil or Martinique, and it is to be hoped that material from these regions may be forthcoming in the near future.

There is another aspect to the problem which has so far not been considered. In the comparison of various Negro groups with one another there is always the question as to whether they are equally of Negro blood. The earlier work of Ferguson (6) found the lighter colored Negroes superior to the darker, and his conclusion was that intelligence increased as the degree of White intermixture increased. The more recent work of Herskovits (11), Peterson and Lanier (19), and Klineberg (15) has not, however, corroborated this finding. In any case, even if the New York group has more White intermixture than the southern Negro group taken as a whole, this would hardly account for the fact that a year-by-year improvement can be noted among the subjects as length of residence in New York increases. In addition, this investigation included two more direct checks of the hypothesis that achievement in the tests might in some way be connected with degree of White intermixture. In the study by Brown (see above) each subject was given, by inspection, a score of from one to five, one meaning exceedingly fair, and five exceedingly black. Table 33 gives the average skin color for the various groups.

TABLE 33

RELATION BETWEEN SKIN COLOR AND LENGTH OF NEW YORK RESIDENCE, TEN-YEAR-OLD BOYS (BROWN)

	Number of Cases	Average Skin Color	Average Mental Age
Less than 2 Years	20	2.95	7.25
2–5 Years	20	3.00	7.65
More than 5 Years	20	3.05	7.50
All Southern born	60	3.00	7.47
Northern born	50	2.92	8.65

It is clear that there is no direct connection between skin color and either length of residence in New York or standing in the tests.

The second check was made on a larger group of subjects and with much more refined techniques by R. E. Holmes (uncompleted thesis for the degree of Master of Arts). He tested 306 of the subjects in Marks' study (see above) for three Negroid characteristics, skin color, lip thickness, and nose width. The skin color was tested by means of the Milton-Bradley color top; the score representing degree of Negro blood is given in terms of percent of black, corrected by the addition of 59 percent of the amount of red (see Herskovits, 12). Table 34 gives the average figures for these three Negroid characteristics.

TABLE 34

RELATION BETWEEN NEGROID CHARACTERISTICS AND LENGTH OF NEW YORK RESIDENCE, TWELVE-YEAR-OLD BOYS (HOLMES)

Group	N	Av. Percent Black	Av. Lip Thickness in mm.	Av. Nose Width in mm.
1 Year	18	78.5	20.7	36.5
2 Years	32	78.0	22.7	32.2
3 Years	16	78.3	24.3	36.2
4 Years	25	80.1	21.9	37.3
5 Years	16	78.2	21.9	38.6
6 Years	34	79.4	21.7	37.2
7 Years	49	77.5	18.6	35.3
8 Years	40	80.4	18.5	35.9
9 Years or more	6	88.8	18.5	35.8
New York born	58	77.4	20.9	36.5

There is again no correspondence whatever between these Negroid characteristics and length of residence in New York, and it is safe to rule out the hypothesis that the amount of White intermixture has very much to do with the results reported in this investigation.

The conclusion remains, therefore, that length of residence in a favorable environment plays an important part in the intellectual level of the Negro children. As for selective migration, it seems to the writer that the use of such a blanket term to cover migration everywhere and at all times is entirely unwarranted. Obviously there is some selection. Not all the members of a given community move, and there must be certain factors which cause some of them to move and others to remain behind. That these factors are, however, necessarily connected with intelligence has so far not been proved. The evidence summarized in an earlier chapter suggested rather that a host of nonintellectual factors might play a part. The problem of selection undoubtedly varies with different communities; factors which operate in one region or in one race may be entirely absent in another; there is no formula which applies to them all. As far as intelligence goes, the material reported in this study gives evidence to the effect that the Negro who leaves the South for the North is not on the average superior to the Negro who remains behind, and that the present superiority of the northern over the southern Negro may be explained by the more favorable environment, rather than by selective migration.

BIBLIOGRAPHY

1. American Academy of Political and Social Science, Annals. Donald Young, editor. Philadelphia, Pa., 1928, Vol. V.
2. Brigham, C. C., A Study of American Intelligence. Princeton, N.J., Princeton University Press, 1920.
3. —— Intelligence Tests of Immigrant Groups. *Psychological Review*, XXXVII (1930), 158-65.
4. Chicago Commission on Race Relations, The Negro in Chicago. Chicago, University of Chicago Press, 1922.
5. Embree, Edwin R., Brown America. New York, The Viking Press, 1931.
6. Ferguson, G. O., The Psychology of the Negro. "Archives of Psychology," 1916, No. 36.
7. Foreman, Clark, Environmental Factors in Negro Elementary Education. New York, W. W. Norton & Co., 1932.
8. Garrett, H. E., Statistics in Psychology and Education. New York, Longmans Green, 1926.
9. Garth, T. R., Race Psychology. New York, McGraw-Hill, 1931.
10. Graham, V. T., Health Studies of Negro Children. "U. S. Public Health Reports," XLI (1926), 2759-83.
11. Herskovits, M. J., On the Relation between Negro-White Mixture and Standing in Intelligence Tests. *Pedagogical Seminary*, XXXIII (1926), 30-42.
12. —— The Anthropometry of the American Negro. New York, Columbia University Press, 1930.
13. Johnson, Charles S., The Negro in American Civilization. Henry Holt, New York, 1930.
14. Kennedy, Louise V., The Negro Peasant Turns Cityward. New York, Columbia University Press, 1930.
15. Klineberg, Otto, An Experimental Study of Speed and Other Factors in "Racial" Differences. "Archives of Psychology," 1928, No. 93.
16. —— A Study of Psychological Differences between "Racial" and National Groups in Europe. "Archives of Psychology," 1931, No. 132.
17. National Academy of Sciences, Memoirs. R. M. Yerkes, Editor. Washington, D.C., XV (1921).
18. Paterson, D. G., and R. M. Elliott, Minnesota Mechanical Ability Tests. Minneapolis, University of Minnesota Press, 1930.

19. Peterson, J., and L. H. Lanier, Studies in the Comparative Abilities of Whites and Negroes. "Mental Measurement Monographs," 1929, No. 5.

20. Phelps-Stokes Fund, Educational Adaptations: Report of Ten Years' Work, 1910-1920.

21. Pintner, R., Intelligence Testing—Methods and Results. New York, Henry Holt, 1931.

22. Pintner, R., and D. G. Paterson, A Scale of Performance Tests. New York, Appleton, 1925.

23. Reuter, E. B., The American Race Problem: a Study of the Negro. New York, T. Y. Crowell Co., 1927.

24. Shimberg, M. E., An Investigation into the Validity of Norms with Special Reference to Urban and Rural Groups. "Archives of Psychology," 1929, No. 104.

25. Wesley, Charles H., Negro Labor in the United States, 1850-1925. New York, The Vanguard Press, 1927.

26. Witty, P. A., and H. C. Lehman, Racial Differences: the Dogma of Superiority. *Journal of Social Psychology*, I (1930), 394-419.

27. Woofter, T. J., and others, Negro Problems in Cities. New York, Doubleday Doran, 1928.

INDEX

Arlitt, A. H., 2
Armstrong Association, 10
Army testing, 1, 2, 37
Atlanta, rural migrants, 54

Barnes, I., 2
Birmingham, migrants, 16ff.
Bloodgood, Annabelle, 15
Boas, Franz, vii
Brigham, C. C., 37, 63
Brown, B. H., 47ff., 60

Carter, F., 15
Charleston, migrants, 19-20
Chicago Defender, 6
Chicago Race Commission, 29, 63
City residence and test score, 53ff.
Collins, S. D., 2
Curtis Arithmetic Tests, 51ff.

Davis, R. A., 2

Elliott, R. M., 63
Embree, E. R., 57, 63

Ferguson, G. O., 60, 63
Foreman, Clark, 58, 63

Garrett, H. E., 63
Garth, T. R., 1, 2, 63
Goodenough, Florence L., 2
Grade retardation, 28ff., 31ff., 34, 36
Graham, V. T., 2, 44, 63

Hand, Frances T., 16, 39
Haynes, George, 10
Herskovits, M. J., 60, 61, 63
Hirsch, N. D. M., 2
Holmes, R. J., 60ff.
Horowitz, E. L., 49ff.

Intelligence: inheritance, 16; city residence, 53ff.; northern residence, 24ff.; tests, sex differences, 39

Johnson, C. S., vii, 57, 58, 63
Jordan, A. H., 2

Keller, R., 2
Kempf, G. A., 2

Kennedy, L. V., 6, 7, 63
King, Louis E., 11
Klineberg, Otto, 60, 63

Lacy, L. D., 2
Lanier, L. H., 3, 12, 24, 29, 44, 60, 64
Lapidus, George, 25ff., 38, 39
Lehman, H. C., 1, 64

Marks, Eli, 34, 40, 61
Mels, Edgar, 8
Migrants: Birmingham, 16ff.; changes in quality, 26ff., 37ff.; Charleston, 19-20; destination, 21; Nashville, 18; rural, 32ff.; 53ff.; school records, 14ff.
Migration: causes, 6, 7, 11; history, 6, 7; newspaper accounts, 7ff.; selective, 4, 12, 53ff.
Milton Bradley color top, 61
Minnesota Paper Form Board, 49ff., 63

Nashville: migrants, 18; rural migrants, 54
National Intelligence Test studies, 25ff., 54
National Urban League, 7
Negro, intelligence tests of, 1, 2, 3, 24ff.
Negroid characteristics, 61
New Orleans, rural migrants, 54

O'Shea, M. V., 2
Otis tests, 41ff.

Paterson, D. G., 63, 64
Performance tests, 47ff.
Peterson, Joseph, vii, 2, 3, 12, 24, 29, 44, 60, 64
Phelps-Stokes Fund, report, 57, 64
Phillips, F. M., 2
Pintner, R., 1, 2, 53, 64
Pintner-Paterson Performance Tests, 47ff.
Pressey, S. L., 2

Reuter, E. B., 29, 64
Roberts, 9
Rogosin, Henry, 45ff.

Schools: expenditures, 57; north and south, 56ff.
Schwegler, R. A., 2

Shimberg, M. E., 53, 64
Skin color, 60
Skladman, Jeannette, 43ff.
Speed tests, 49
Stanford-Binet tests, 43ff.
Strachan, L., 2
Sunne, Dagny, 2

Teter, G. F., 2
Thibodaux, 54
Traver, I. D., 41ff.

Wallach, Elsie, 44ff.
Washington, F. B., 10
Wesley, C. H., 7, 64
Winn, E., 2
Witty, P. A., 1, 64
Woofter, T. J., 10, 56, 58, 64
Work, M. C., 5

Yates, Charlotte, 28, 30ff., 39, 40
Yerkes, R. M., 63
Young, Donald, 63

COLUMBIA UNIVERSITY PRESS
COLUMBIA UNIVERSITY
NEW YORK

———

FOREIGN AGENT
OXFORD UNIVERSITY PRESS
HUMPHREY MILFORD
AMEN HOUSE, LONDON, E.C. 4

Bei Fragen zur Produktsicherheit wenden Sie sich bitte an:
If you have any questions regarding product safety,
please contact:

Walter de Gruyter GmbH
Genthiner Straße 13
10785 Berlin
productsafety@degruyterbrill.com